SHEET METAL FORMING PROCESSES AND DIE DESIGN

Second Edition

SHEET METAL FORMING PROCESSES AND DIE DESIGN

Second Edition

VUKOTA BOLJANOVIC, Ph.D.

Industrial Press

A full catalog record for this book is available from the Library of Congress.
ISBN 978-0-8311-3492-1

Industrial Press, Inc.
32 Haviland Street, Unit 2C
South Norwalk, CT 06854

Sponsoring Editor: John Carleo
Developmental Editor: Robert Weinstein
Interior Text and Cover Design: Janet Romano

10   9   8   7   6   5   4   3   2   1

DEDICATION

To my granddaughter *Milla Liv Broadwater*
and
to my grandson *Alexandar Boljanovic*

TABLE OF CONTENTS

PREFACE TO THE SECOND EDITION

The science, engineering, and technology of sheet metal forming processes and die design continue to advance rapidly on a global scale and with major impact on the economies of all nations. In preparing this second edition, my goal throughout has been to provide an expanded and more comprehensive treatment of the sheet metal forming processes, while placing forming processes and die design in the broader context of the techniques of press-working sheet metal.

As in the first edition, this text presents topics with a balanced coverage of relevant fundamentals and real-world practices to help the reader develop an understanding of the often complex plastic deformation process. The book is now a more complete text for academics as well as for practicing designers and manufacturing and industrial engineers.

This new edition has essentially the same introductory nature, format, and organization as the first edition. However, enhancements and extensions of several of the chapters in the First Edition have been included, and one new chapter on a topic of great importance to the modern manufacturing industry has been added.

What Is New in this Edition

A careful page-by page comparison with the first edition will show that hundred of changes have been made to improve the clarity and readability of the numerous topics covered.

- Chapter 1 (concerning the structure of metals) has been rewritten and sections on atomic structure and bonding between atoms and molecules have been added.
- Chapter 2 (concerning mechanical behavior of metals) has new sections on hardness and the effect of temperature on material properties.
- Chapter 4 (concerning blanking and punching) has new examples on blanking clearance and force.
- Chapter 5 (concerning bending) has a rewritten section on bending force as well as new examples.
- Chapter 6 (concerning deep drawing) has a rewritten section on blank calculation for rotational symmetrical shells and new methods for blanking calculation.
- Chapter 7 (concerning various forming processes) has a rewritten section on flexible die forming, while new sections discussing the Guerin process, Verson-Wheelon process, Marform process, and hydroforming process have been added.
- All of the illustrations in the book have been redrawn for improved graphic impact and clarity, and numerous new illustrations have been added.
- A new Chapter 14 (on the topic of quick die-change systems and die design) has been added, as has an Appendix 4 on the topic of technical specifications of helical and Belleville springs.
- The glossary and bibliography at the end of the book have been thoroughly updated.

I would like to thank my colleagues and readers of the book for their helpful suggestions regarding improvements to the second edition, and my family and personal friends who indirectly contributed to this book through their love and friendship during the last few years.

Vukota Boljanovic

PREFACE TO THE FIRST EDITION

A very large variety of sheet-metal forming processes is used in modern sheet-metal press-working shop practice. Many of these deformation processes, used in making aircraft, automobiles, and other products, use complex equipment that is derived from the latest discoveries in science and technology. With the ever-increasing knowledge of science and technology, future deformation processes promise to be even more complex to satisfy the demand for more productivity, lower cost, and greater precision. However, for all their advantages, the more sophisticated deformation processes of today have not replaced the need for basic sheet-metal forming processes and dies.

This book draws on the author's 30-plus years of experience as an engineer and provides a complete guide to modern sheet-metal forming processes and die design — still the most commonly used manufacturing methodology for the mass production of complex, high-precision parts. Much more practical than theoretical, the book covers the "hows" and "whys" of product analysis, and the mechanisms of blanking, punching, bending, deep drawing, stretching, material economy, strip design, movement of metal during stamping, and tooling design. Readers will find numerous illustrations, tables, and charts to aid in die design and manufacturing processes; Formulas and calculations needed for various die operations and performance evaluation are included; and designations, characteristics, and typical applications of various carbon and alloy steels for different die components are evaluated.

The book concentrates on simple, practical engineering methods rather than complex numerical techniques to provide the practicing engineer, student, technician, and die maker with usable approaches to sheet-metal forming processes and die design.

The first part of the book deals with the structures of metals and the fundamental aspects of the mechanical behavior of metals. Knowledge of structures is necessary to controlling and predicting the behavior and performance of metals in sheet-metal forming processes.

The second part of the book covers all aspects of forming sheet metal. It presents the fundamental sheet-metal forming operations of shearing, blanking and punching, bending, stretching, and deep drawing. Mechanics of various drawing processes indicate ways in which the deformation, loads, and process limits can be calculated for press forming and deep drawing operations. The book includes various drawing processes (nosing, expanding, dimpling, spinning and flexible die forming) mostly used in the aircraft and aerospace industry.

Dies are very important to the overall mass production picture, so they are discussed in the last section of the book, which presents a complete picture of the knowledge and skills needs for the effective design of dies for sheet-metal forming processes described. Special attention is given to:

- Formulas and calculations needed for various die parts.
- Rules of thumb and innovative approaches to the subject.
- Properties and typical applications of selected tool and die materials for various die parts.

Although the book provides many examples of calculations, illustrations, and tables to aid in sheet-metal forming processes, die design, and die manufacturing, it should be evident that it is not possible to present all the data, tables, statistics, and other information needed to design complicated dies and other tools for sheet-metal forming in one text. However, the book endeavors to provide most of the information needed by a die designer in practical situations.

The author owes much to many people. No book can be written in a vacuum. I am grateful to my wife, who understands my need for long periods of isolation. I also wish to express my deepest appreciation and thanks for the competent work of Em Turner Chitty, who labored countless hours at editing and proofreading. Finally, I wish to thank my English language teacher from The University of Tennessee, Anwar F. Accawi, who encouraged me to begin writing this book.

Vukota Boljanovic
Knoxville, Tennessee

LIST OF TABLES

INTRODUCTION

I.1 BASIC CHARACTERISTICS OF SHEET METAL FORMING PROCESSES

Sheet metal parts are usually made by forming material under cold conditions. (In many cases, however, sheet-metal parts are formed under hot conditions because the material, when heated, has a lower resistance to deformation—i.e., it is more easily deformed).

Strips or blanks are very often used as initial materials and are formed on presses using appropriate tools. The shape of a part generally corresponds to the shape of the tool.

Sheet-metal forming processes are used for both serial and mass production. Their characteristics include high productivity, highly efficient use of material, easy servicing of machines, the ability to employ workers with relatively lower basic skills, and other advantageous economic aspects. Parts made from sheet metal have many attractive qualities: good accuracy of dimension, adequate strength, light weight, and a broad range of possible dimensions, from the miniature parts used in electronics to the large components of airplane structures.

I.2 CATEGORIES OF SHEET METAL FORMING PROCESSES

All sheet-metal forming processes can be divided into two groups:

1. Cutting processes: shearing, blanking, punching, notching, cutoff, shaving, trimming, parting, slotting and perforation, and lancing.
2. Plastic deformation processes: bending, twisting, curling, deep drawing, spinning, stretch flanging, shrink flanging, hole flanging, necking, bulging, ribbing, hemming, and seaming.

a) The First Group of Processes

Cutting processes involve cutting material by subjecting it to shear stresses usually between punch and die or between the blades of a shear. The punch and die may be any shape, and the cutting contour may be open or closed.

Shearing. Shearing involves the cutting of flat material forms from sheet, plate, or strip. The process may be classified by the type of blade or cutter used, whether straight or rotary.

Blanking. Blanking involves cutting the material to a closed contour by subjecting it to shear stresses between punch and die. In this process, the slug is usually the work part and the remainder is scrap.

Punching. Punching is the cutting operation by which various shaped holes are sheared in blanks. In punching, the sheared slug is discarded, and the material that surrounds the punch is the component produced.

Notching. Notching is cutting the edge of the blank to form a notch in the shape of a portion of the punch. If the material is cut around a closed contour to free the sheet metal for drawing or forming, this operation is called semi-notching.

Cutoff. Cutoff is a shearing operation in which the shearing action must be along a line. A cutoff is made by one or more single line cuts.

Shaving. Shaving is a cutting operation that improves the quality and accuracy of blanked parts by removing a thin strip of metal along the edges. Only about 100 microns (0.004 inches) of material are removed by shaving.

Trimming. For many types of drawing and forming parts, excess metal must be allowed; the workpiece can then be held during the operation that shapes the metal into the form of the part. The cutting off of this excess metal after drawing or forming operations is known as trimming.

Parting. Parting is the cutting of a sheet metal strip by die cutting edges on two opposite sides. During parting, some amount of scrap is produced. This might be required when the blank outline is not a regular shape and cannot perfectly nest on the strip.

Lancing. Lancing is an operation in which a single line cut is made partway across the work material. No material is removed, so there is no scrap

b) The Second Group of Processes

Involves partial or complete plastic deformation of the work material.

Bending . Bending consists of uniformly straining flat sheets or strips of metal around a linear axis. Metal on the outside of the bend is stressed in tension beyond the elastic limit. Metal on the inside of the bend is compressed.

Twisting. Twisting is the process of straining flat strips of metal around a longitudinal axis.

Curling. Curling is a forming operation that is used to strengthen the edge of sheet metal. The edges of the drawn part are formed into a roll or curl for providing a smooth rounded edge.

Deep Drawing. Deep drawing is a sheet metal forming process in which a sheet metal blank is radially drawn into a cylindrical or box-shape forming die by the mechanical action of a punch. It is thus a shape transformation process with material retention. Drawing may be performed with or without a reduction in the thickness of the material.

Spinning. Spinning is a process of forming the workpiece from a circular blank or from a length of tubing. All parts produced by spinning are symmetrical about the central axis.

Stretch flanging. During the forming of the flange, the periphery is subjected to tension. Stretched flanges are susceptible to splitting and thinning. Cracking or tearing of the metal is the most common defect in this type of flange.

Shrink flanging. The flange is subjected to compressive stress, and it is susceptible to wrinkling and thickening.

Hole flanging. Hole flanging is the operation of forming the inner edges in a punched or drilled hole. Stretching the metal around the hole subjects the edges to high tensile strains, which could lead to cracking and tearing the edges.

Necking. Necking is an operation by which the top of a cup may be made smaller than its body.

Bulging. Bulging is a process that involves placing a tubular, conical, or curvilinear part in a split female die and expanding it with pressure from the polyurethane punch.

Ribbing. To increase the rigidity of sheet metal formed parts, small ribs can be formed between bent or formed flanges on the remainder of the flat surface of the part. This rib is formed directly from the sheet metal itself.

Hemming. Hemming is a bending operation that bends sheet metal edge over on itself in more one bending step. This is often used to remove ragged edges and to improve the rigidity of the edge.

Seaming. Seaming is an operation that is used to make stovepipes, food cans, drums, barrels, and other parts that are made from flat sheet metal.

I.3 CHARACTERISTICS OF SHEET METAL PARTS AND THEIR TECHNICAL PROPERTIES

The designers of products made from sheet metal have a huge responsibility (and legal liability) for inventing the exact design that will result in optimum production considering the complexity of the technological factors, the kind and number of operations, the production equipment (machine and tools required), the material expenses, and the quantity and quality of material.

It is necessary to design technical components and operations so as to fulfill the product specifications optimally. The design of a part is adequate if it combines the most economical production with the most satisfactory quality.

To arrive at the best and most economical product, the following parameters must be observed: a process resulting in minimum production of scrap, using standard equipment and machines wherever possible, the minimum possible number of operations, and relatively lower-skilled workers. Generally speaking, the most important of these factors is cost. That is the most efficient design should also have the lowest possible cost.

To ensure high quality production, and lower the cost of the product, it is necessary to abide by some basic recommendations, such as using a minimal drawing radius, a minimal bending radius, and minimal dimensions of punch holes depending on the material thickness. The method used to dimension a drawing is also very important and has a great influence on the quality and price of a part.

Part
One

Theoretical Fundamentals

In this part, the basic theory of the plastic deformation of metals is discussed. The material is divided into two chapters that focus respectively on the structure of metals and the mechanical behavior of metals. For solving many problems, the so-called "engineer's method" is used. In consideration of the intended audience for this book, other methods such as Slip Line Analysis and Full Energy Deformation are not explained; they are more important in research. For practical applications, the "engineer's method" is more useful.

1

One

The Structure of Metals

1.1 INTRODUCTION

Metals have played a major role in the development of civilization. Why have metals been useful to humanity through the ages? Why do metals have a characteristic metallic shine? Why are some metals solid? Why are some metals malleable and ductile? Why are metals good conductors? Why are some metals soft and others hard? The answers to these and similar questions can be provided by gaining an understanding of the atomic structure and metallic bond of the metal atoms to form a crystalline structure. In general, processing is used to control the structures of metals, and the structure determines the properties of the material. One of the principal ways of controlling the structure of materials is to use processing techniques to manipulate the phases that are present in the material and how these phases are distributed. The main activity of metallurgists and many other materials scientists is to manipulate the structure of materials to control their properties. This manipulation can be used to control and predict the behavior and performance of metals in various manufacturing processes. That is to say, it is vitally important to understand the structure of metals in order to predict and evaluate their properties.

1.2 ATOMIC STRUCTURE

What we call matter is that which has mass and takes up space. The basic structural unit of matter is the atom. The word *atom* is derived from the Greek word *atom*, which means "indivisible." Accordingly, atoms cannot be chemically subdivided by ordinary means.

Atoms are composed of three types of particles: protons, neutrons, and electrons (Fig. 1.1). Protons have a positive (+) charge, surrounded by enough negatively charged (−) electrons so that the charges are balanced. The number of electrons in an atom identifies both the atomic number and the element itself. Neutrons, as is indicated by their name, have no charge.

Protons and neutrons are responsible for most of the mass in an atom. The mass of an electron is very small (9.108×10^{-28} grams). There are 103 elements; they serve as the chemical building blocks of all matter. In principle, each element can exist as a solid, liquid, or gas, depending on the temperature and pressure conditions under which it exists at any given moment. The same element may also contain varying numbers of neutrons; this variation in the number of neutrons creates different forms of an element that are called *isotopes*.

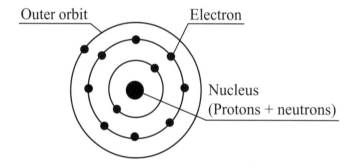

Fig. 1.1 Simple model of atomic structure.

The properties of matter depend on which atoms are involved in a material and how they are bonded together.

Metals behave differently from ceramics, and ceramics behave differently from polymers. The structures of materials are classified according to the general magnitude of the various features being considered. The three most common major classifications of structural are

- Atomic structure, which includes features that cannot be seen, such as the types of bonding between the atoms and the way the atoms are arranged.
- Microstructure, which includes features that can be seen using a microscope, but seldom with the naked eye.
- Macrostructure, which includes features that can be seen with the naked eye.

The atomic structure primarily affects the chemical, physical, thermal, electrical, magnetic, and optical properties of a material. The microstructure and macrostructure can also affect these properties but they generally have a greater effect on mechanical properties and on the rate of chemical reaction. The properties of a material offer clues to its structure. For example, if a metal is strong, this suggests that its atoms are held together by strong bonds. However, even strong bonds must allow atoms to move, because even strong metals are usually formable. To understand the structure of a material, it is necessary to know the type of atoms present and how the atoms are arranged and bonded.

1.3 BONDING BETWEEN ATOMS AND MOLECULES

Atoms are rarely found as free and independent units; instead, they are held together in molecules by various type of bonds as a result of interatomic forces.

It is known that the atomic structure of any element is made up of a positively charged nucleus (protons plus neutrons) surrounded by electrons revolving around it. The atomic number of an element indicates the number of positively charged protons in its nucleus. To determine the number of neutrons in an atom, the atomic number is simply subtracted from the atomic weight.

The electric structure of the atoms influences the nature of the bond, which may be stronger (primary) or weaker (secondary).

1.3.1 Primary Bonds

Primary bonds include the following forms:

- ionic
- covalent
- metallic

Ionic and covalent bonds are called *intramolecular* bonds because they involve attractive forces between atoms within the molecules.

a) Ionic Bond

When metallic and nonmetallic atoms come together, an ionic bonding occurs. Metals usually have 1, 2, or 3 electrons in their valence (outer) shell. Nonmetals have 5, 6, or 7 electrons in their valence shell. Atoms with outer shells that are only partially filled are unstable. To become stable, a metal atom "wants" to get rid of one or more electrons in its outer shell. Losing electrons will either result in an empty outer shell or get it closer to having an empty outer shell. The atom "wants" to have an empty outer shell because the next lower energy shell will have eight electrons, which will make the molecule stable. (In general, eight electrons in the outermost shell is the most stable atomic configuration, and nature provides a very strong bond between atoms that achieve this configuration.) Figure 1.2 illustrates the process for an ionic bond between sodium (Na) and chlorine (Cl) atoms, the combination that yields common salt.

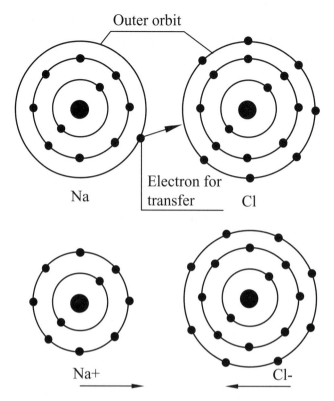

Fig. 1.2 Schematic illustration of ionic bonding.

Sodium has one valence electron that it "wants" to shed so that it can become stable with a full outer shell of eight. Its counterpart, chlorine, has seven valence electrons, so it "wants" to gain an electron in order also to have a full shell of eight. All of the positively charged atoms attach to all of the negatively charged atoms. For example, sodium ions surround themselves with negative chlorine ions, and chlorine ions surround themselves with positive sodium ions. The attraction is equal in all directions and results in a three-dimensional structure rather than the simple link of a single bond.

Materials joined by ionic bonds have the following characteristics:

- High hardiness, because particles cannot easily slide past one another.
- Good electrical insulative properties, because all charge transport must come from the ions' movement, but there are no free electrons or ions, so there is no transport for electrical charges.
- High melting point because ionic bonds are relatively strong.
- Moderate to high strength.

b) Covalent Bonds

A bond between two nonmetal atoms is usually a covalent bond. Nonmetals have four or more electrons in their outer shells. With so many electrons in the outer shell, it would require more energy to remove the electrons than would be gained by making new bonds. Thus, in the covalent bond, electrons are shared between atoms so that each achieves a stable electron set of eight in its outermost shell. The shared negative electrons are located between the positive nuclei to form the bonding link. Figure 1.3 illustrates this type of bond for chlorine, where two atoms, each containing seven electrons in its outer shell, share a pair to form a stable molecule.

More than one electron pair can be formed, with half of the electrons coming from one atom and the rest from the other atom. An important feature of this bond is that the electrons are tightly held and equally shared by the participating atoms. The atoms can be of the same element or different elements. In each molecule, the bonds between the atoms are strong, but the bonds between the molecules on the whole are usually weak.

Because most covalent compounds contain only a few atoms and because the forces between molecules are weak, most covalent compounds have low melting and boiling points. However, in some, like diamond, which is carbon, each atom has four neighbors with which it shares electrons. This produces a very rigid three-dimensional structure, accounting for the extremely high hardness of this material. Engineering materials possessing covalent bonds tend to be polymeric.

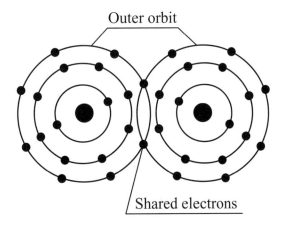

Fig. 1.3 Schematic illustration of a covalent bonding.

c) Metallic Bond

When the atoms of pure metals or metal alloys bond, a metallic bond occurs. Generally, metallic elements possess only one to three electrons in the other shell and the bond between these electrons and the nucleus is relatively weak, whereas the remainder are held firmly to the nucleus. The result is a structure of positive ions (nucleus and nonvalent electrons) surrounded by a wandering assortment of universally shared valence electrons (electron cloud), as seen in Fig. 1.4.

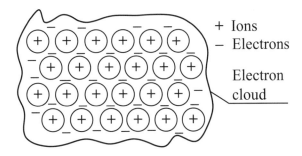

Fig. 1.4 Schematic illustration of metallic bond.

In this arrangement, the valence electrons have considerable mobility and are able to conduct heat and electricity easily. Also, the non-localized nature of the bonds makes it possible for the atoms to slide past each other when the metal is deformed or shaped. Moreover, they provide "cement" necessary to produce the positive-negative-positive attractions necessary for bonding.

Materials bonded by metallic bonds have the crystalline structure of metals that can be deformed by atom movement mechanisms. This is the basis of metal plasticity, ductility, and many of the shaping processes used in metal fabrication.

1.3.2 Secondary Bonds

Weak or secondary bonds involve attraction forces between molecules or intermolecular forces. There is no transfer or sharing of electrons in secondary bonding. They are found in most materials, but their effects are often overshadowed by the strength of the primary bonding. There are two types of secondary bonds:

- Van der Waals bonds
- hydrogen bonds

a) Van der Waals Bond

The bonds between the molecules that allow sliding and rupture to occur are called van der Waal forces. They are usually formed when an uneven charge distribution occurs, creating what is known as a dipole. The total charge is zero, but there is slightly more positive or negative charge on one end of the atom than on the other (an effect referred to as *polarization*). Some molecules, such as hydrogen and water, can be more positive or negative than others. The negative part of one molecule tends to attract the positive part of another to form a weak bond.

Another weak bond can result from the rapid motion of the electrons in orbit around the molecule; temporary dipoles form when more electrons happen to be on one side of the molecule than the other. This instantaneous polarization provides forces of attraction between molecules called the *dispersion effect*. Van der Waals bonds occur to some extent in many materials, but are particularly important in plastics and polymers.

b) Hydrogen Bond

The hydrogen bond is a special case of dipole forces. A hydrogen bond is the attractive force between the hydrogen, which has positive partial charge when it is attached to an electronegative atom of one molecule, and an electronegative atom of a different molecule. Usually the electronegative atom is oxygen or nitrogen, both of which have a partial negative charge.

A hydrogen bond is about 10 percent stronger than a van der Waals bond, but it is still strong enough to have many important ramifications for the properties of water. Such bonds play a significant role in biological systems, but, as with all secondary bonds, are rarely significant for engineering.

1.4 THE CRYSTAL STRUCTURE OF METALS

The key characteristic that distinguishes the structure of metals from that of non-metals is their atomic architecture. When metals solidify from a molten state, the atoms arrange themselves into a crystalline order, according to the number and types of imperfections found in the structure, and to the bonding forces that keep the collection or structure of atoms bound or joined together. Metallic materials have electrons (ions) that are free to move about their positive centers, sitting in an electron cloud or gas, which acts to bond the ions together. The existence of these free electrons and their movement, within limits, has a number of consequences for the properties of metals. For example, the slight movement of the atoms under the influence of an external load is called elastic deformation. One of the most important characteristics of metals is that freely moving electrons can conduct electricity, so that metals tend to be good electrical and thermal conductors.

In a crystalline structure the atoms are arranged in a three-dimensional array called a lattice. The lattice has a regular repeating configuration in all directions resulting from the forces of chemical bonding. The repeated pattern controls properties such as strength, ductility, density, conductivity, and shape.

Most metals only exist in one of the three common basic crystalline structures:

- body-centered cubic (bcc)
- face-centered cubic (fcc)
- and hexagonal close packed (hcp)

These structures are shown in Fig. 1.5. Each point in this illustration represents an atom. The distance between the atoms in these crystal structures, measured along one axis, is known as the constant or parameter of the lattice.

As shown in Fig. 1.5a, a body-centered cubic (bcc) structure has an atom at each corner of an imaginary cube and an atom in the center of this cube (8+1). A (bcc) structure generally produces strong metals, which are reasonably ductile. Examples of (bcc) metals are chromium, alpha iron, and molybdenum.

Face-centered cubic (fcc) structures, as shown in Fig. 1.5b, have an atom at the corners of a cube and an atom at the center of each face of the cube (8+6). Metals with (fcc) structures tend to be soft and ductile over a wide range of temperatures. Examples of (fcc) metals are aluminum, copper, and nickel.

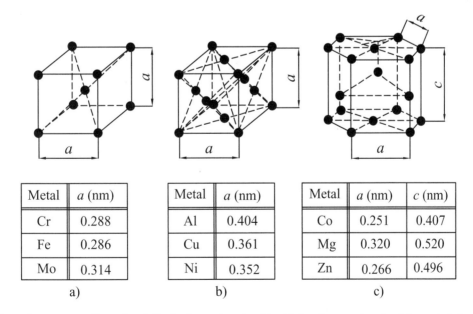

Metal	a (nm)
Cr	0.288
Fe	0.286
Mo	0.314

a)

Metal	a (nm)
Al	0.404
Cu	0.361
Ni	0.352

b)

Metal	a (nm)	c (nm)
Co	0.251	0.407
Mg	0.320	0.520
Zn	0.266	0.496

c)

Fig. 1.5 Crystal structure of metals: a) the body-centered cubic; b) the face-centered cubic; and c) the hexagonal close-packed.

Hexagonal closed-packed (hcp) structures (Fig 1.5c) have an atom at the corners of an imaginary hexagonal prism and an atom at the center of each hexagonal face, and a triangle of atoms in between the hexagons, which rest in the shaded valleys (12+2+3). These metals, such as cobalt, magnesium, and zinc, are relatively brittle.

In the cubic structures, the parameter has the same dimensions in all three ortho-axes. However, the hexagonal close-packed structure has different dimensions.

The properties of a particular metal are very dependent on its crystal structure. Some metals can exist in more than one crystal structure at different temperatures. Iron and steel, for example, can exist in both (bcc) and (fcc) structures. Other examples of allotropic metals, as they are called, are titanium (hcp and bcc) and iron. Interestingly, iron has a (bcc) structure below 900 and at 1400 , while between those temperatures, iron changes its structure to (fcc), which requires less energy to maintain. Above 1400 iron reverts to its original structure.

The appearance of more than one type of crystalline structure is known as allotropism or polymorphism, meaning "many shapes." Allotropism is an important aspect in the heat treatment of metals and welding operations.

The atoms in crystals cannot move far from their place in the cubic structure but they swing around their balance position. The planes in crystal structures that contain most atoms are known as netting planes.

The basic crystallographic planes in cubic lattice structures are these:

- the plane of a cube with index 100 (Fig. 1.6a)
- the plane of a rhombus dodecahedron with index 110 (Fig. 1.6b)
- the plane of an octahedron with index 111 (Fig. 1.6c)

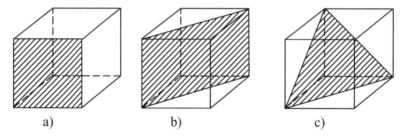

a)   b)   c)

Fig. 1.6 The basic crystallographic planes in cubic lattice structures: a) the plane of a cube with index 100, b) the plane of a rhombus dodecahedron with index 110, c) the plane of an octahedron with index 111.

The properties of metals (physical, chemical, and mechanical) depend on the arrangement of the crystal lattice structure and the magnitude of the distance between the atoms, so the characteristics in different directions will be different. This rule applies, for example, to the modulus of elasticity (the ratio of stress to strain in the elastic ranges, in tension or compression). The modulus of elasticity, measured in different directions in the same metal (across or along the direction of rolling, for example), is different. The measurements are performed on the three main crystallographic planes as described above. The results are given in Table 1.1.

Table 1.1 Value of the modulus of elasticity, measured in different directions

METAL	Modulus of elasticity E (GPa)		
	(100)	**(110)**	**(111)**
Aluminum	63.7	72.6	76.1
Copper	66.7	130.3	191.1
Iron	125.0	210.5	272.7
Tungsten	384.6	384.6	384.6

1.5 DEFORMATION AND STRENGTH OF SINGLE CRYSTALS

The electrons in the metallic bond are free to move about their positive ions in an electron cloud that acts to glue or bond the ions together. This free movement, within limits, also allows for the movement of the atoms under the influence of external forces. This slight movement (visible only under the most powerful microscopes) is called *elastic deformation* or *elastic strain*. After an external force such as a bending force is removed, the internal electrical force that causes the atoms to move will decrease, allowing the atoms to return to their normal position; they leave no sign of ever having been moved. An analogy to this type of behavior is a leaf spring that bends when loaded and returns to its original shape when the load is removed.

If too much external force is applied by excessive bending of the leaf spring, the atoms might move too far from their original positions to be able to move back again when the external force is removed. This permanent deformation is known as plastic deformation.

In plastic deformation of metals, there are two mechanisms: slipping and twinning. Slipping (shown in Fig.1.7) occurs when two groups of crystals under a shearing force move in a parallel fashion along what is called the slip plane, causing the two groups to slip so that they are dislodged one or more crystal's length (*a*). This effect is called shear stress.

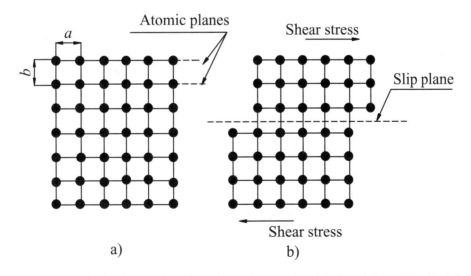

Fig. 1.7 Plastic deformation of a single crystal subjected to a shear stress: a) before deformation; b) deformation by slip.

Twinning (shown in Fig.1.8) occurs when horizontal planes of atoms move proportionally to their distance from the twinning plane. Twinning usually occurs if a plastic deformation is made by striking, whereas slipping occurs more often when plastic deformation occurs as a result of the application of a static load.

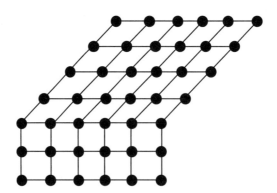

Fig. 1.8 Plastic deformation across the plane of twinning.

In plastic deformation, inter-atomic forces keep their elastic characteristics, which is why plastic deformation is always accompanied by some elasticity. When the external force is removed in plastic deformation, a new arrangement of atoms is established, but the pre-existing distance among them is retained. This statement is valid providing that no stress remains.

In accordance with this explanation, we can conclude that in *elastic* deformations, only the distance between atoms changes, but in plastic deformations, both the distance and the positions of the atoms change.

1.5.1 Imperfections in the Crystal Structure of Metals

Attempts have been made to predict the plastic and mechanical properties of metals by theoretical calculation. With the known value of the inter-atomic forces in the crystal structure of a given lattice, calculations on the theoretical value of the mechanical properties of metals have been made. However, the calculated values of mechanical properties of the metals were much larger than the values obtained with tensile testing in laboratory settings. The actual strength of materials is much lower than the levels obtained from theoretical calculations. This discrepancy has been explained in terms of imperfections in the crystal structure. Unlike the idealized models, metal crystals contain large numbers of defects. Idealized models of crystal structures are known as perfect lattices. Figure 1.9 shows an idealized model of a metal structure without dislocations, in which the atoms from one row correspond with the atoms in another row.

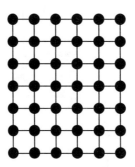

Fig. 1.9 Idealized models of crystal structures.

Orderliness between atoms is not disturbed. Actual metal crystals contain large numbers of defects and imperfections, which can be classified in four categories: point defects, line defects, planar imperfections, and volume imperfections.

a) *Point (or atomic) defects* (Fig. 1.10) can be divided into three main defect categories: the term can mean missing atoms (a vacancy), an interstitial atom (an extra atom) in the lattice, or a foreign atom (substitutional) that has replaced the atom of the pure metal.

Vacancy defects result from a missing atom in a lattice position, and may result from imperfect packing in the process of crystallization, or it may be due to increased thermal vibrations of the atoms brought about by elevated temperatures.

Substitutional defects result from an impurity that is present in a lattice position. Interstitial defects result from an impurity located at an interstitial site or from one of the lattice atoms being in an interstitial position instead of being in its usual lattice position.

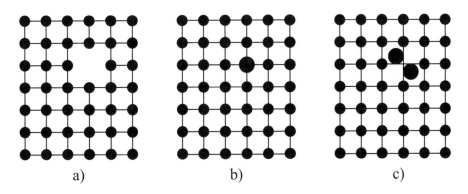

a)   b)   c)

Fig. 1.10 Point defects : a) vacancy defect, b) substitutional defect, c) interstitial defect.

b) *Line defects*, called dislocations, are defects in the orderly arrangement of a metal's atomic structure. Dislocations can be of three types: edge, screw, or mixed, depending on how they distort the lattice. It is important to note that dislocations cannot end inside a crystal. They must end at a crystal edge or other dislocation, or they must close back on themselves.

Figure 1.11 shows a model of a metal structure with edge dislocation. Edge dislocations consist of an extra row or plane of atoms in the crystal structure. The imperfection may extend in a straight line all the way through the crystal or it may follow an irregular path. It may also be short, extending only a small distance into the crystal, causing a slip of one atomic distance along the glide plane (the direction in which the edge imperfection is moving). The slip occurs when the crystal is subjected to a stress, and the dislocation moves through the crystal until it reaches the edge or is arrested by another dislocation.

As can be seen, one column from the upper part does not exist in the lower part of the figure. In this zone, the atoms do not appear in regular order; they are moved or dislocated.

That is why one excess vertical plane must appear in this zone, which functions as an extra plane. Obviously, the atoms near the dislocation in one zone are in a pressured condition, and in another zone they are in a stretched condition. In these conditions, a lattice structure ought to bend. However, if a dislocation is locked with huge atoms, it will not bend, and a dislocation will occur in the place where there is most concentration of strain. A screw dislocation can be produced by a tearing of the crystal parallel to the slip direction. If a screw dislocation is followed all the way around a complete circuit, it will show a slip pattern similar to that of a screw thread.

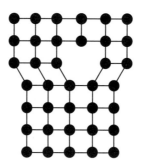

Fig. 1.11 Line defects (dislocations).

c) *Planar imperfections* are larger than line imperfections and occur over a two- dimensional area. Planar imperfections exist at an angle between any two faces of a crystal or crystal form. These imperfections are found at free surfaces, domain boundaries, grain boundaries, or interphase boundaries. Free surfaces are interfaces between gases and solids. Domain boundaries refer to interfaces where electronic structures are different on either side, causing each side to act differently although the same atomic arrangements exist on both sides. Grain boundaries exist between crystals of similar lattice structure that possess different spacial orientations. Polycrystalline materials are made up of many grains that are separated by distances (typically) of several atomic diameters. Finally, interphase boundaries exist between the regions where materials exist in different phases (i.e., bcc next to fcc structures).

d) *Volume imperfections* are three-dimensional macroscopic defects. They generally occur on a much larger scale than the microscopic defects. These macroscopic defects generally are introduced into a material during refinement from its raw state or during fabricating processes.

The most common volume defect arises from foreign particles being included in the prime material. These second-phase particles, called inclusions, are seldom wanted because they significantly alter the structural properties. An example of an inclusion may be oxide particles in a pure metal or a bit of clay in a glass structure.

1.5.2 Grain Size and Boundary

When molten metal begins to solidify, crystals begin to form independently of each other at the various locations within the melted mass. Figure 1.12 is a schematic illustration of the various stages during the solidification of molten metal.

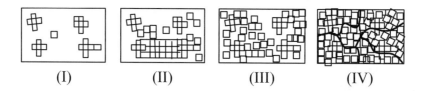

      (I)   (II)   (III)   (IV)

Fig.1.12 Schematic illustration of the various stages during the solidification of molten metal.

Each small square represents a unit cell. The first (I) represents the nucleation of crystals at random sites in the molten metal. The second and third (II and III) represent the growth of crystals as solidification continues. The last (IV) represents solidified metal, showing individual grains and grain boundaries.

The *grain boundary* refers to the outside area of a grain that separates it from the other grains. The grain boundary is a region of misfit between the grains and is usually one to three atom diameters wide. The grain boundaries separate variously oriented crystal regions (polycrystalline) in which the crystal structures are identical. Figure 1.12 (IV) represents five grains of different orientation and the grain boundaries that arise at the interfaces between the grains.

A very important feature of a metal is the average size of the grain. The size of the grain determines the properties of the metal. For example, a smaller grain size increases tensile strength and tends to increase ductility. A larger grain size is preferred for improved high-temperature creep properties. *Creep* is the permanent deformation that increases with time under constant load or stress. Creep becomes progressively easier (greater) with increasing temperature.

Another important property of the grains is their orientation. In general, grains may be oriented in a random arrangement or in a preferred orientation. When the properties of a material vary with different crystallographic orientations, the material is said to be *anisotropic.*

Alternately, when the properties of a material are the same in all directions, the material is said to be *isotropic.* For many polycrystalline materials the grain orientations are random before any deformation of the material is done. When a material is formed, the grains are usually distorted and elongated in one or more directions, which makes the material anisotropic.

In some applications, it is desirable to optimize a material's property in a particular direction. Some manufacturing processes can give a material texture; this means the grains are oriented in one direction instead of randomly. This is called preferred orientation.

1.5.3 Strain Hardening

Movement of an edge dislocation across a crystal lattice under a shear stress does not happen all at the same time; instead, only the local domain slips. Figure 1.13a schematically illustrates a crystal lattice non- deformed metal, and Figs. 1.13b to 1.13d illustrate the movement of an edge dislocation across the crystal lattice under a shear stress. Dislocation helps explain why the actual strength of metals is much lower than that predicted by theory. A slip plane containing a dislocation requires less shear stress to cause slip than a plane in a perfect lattice.

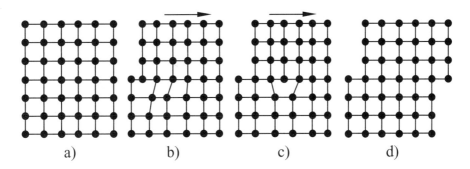

a)   b)   c)   d)

Fig. 1.13 Gradual movement dislocation through a crystal lattice during plastic deformation: a) structure before deformation; b), c), and d) movement of an edge dislocation across the crystal lattice under a shear stress.

Although the presence of a dislocation lowers the shear stress required to cause slip, dislocations can become entangled and interfere with each other and be impeded by barriers, such as grain boundaries and impurities and inclusions in the metal. Entanglement and impediments increase the shear stress required for slip. The increase in shear, and hence the increase in the overall strength of the metal, is known as work hardening or strain hardening. Work hardening is used extensively in strengthening metals in metalworking processes at ambient temperatures.

The relationship between a metal's strength and the number of dislocations in the metal's crystal lattice is shown in Fig. 1.14.

The curve of function UTS = f(n) graphically illustrates an hypothesis regarding strain hardening in metals in metalworking processes at ambient temperature. The steep part of the curve UTS = f(n) for $n < n_1$ characterizes the beginning of the first dislocations and other defects in structure, such as vacancies, interstitial atoms, and impurities. The right part of the curve for $n > n_1$ characterizes a gradual growth of strength and represents aggregate defects.

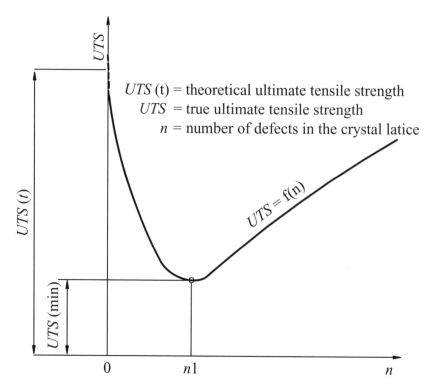

Fig. 1.14 The relationship between the metal's strength and number of dislocations in the metal's crystal lattice.

1.6 RECOVERY AND RECRYSTALLIZATION

In studies of the phenomenon of work hardening, it has been shown that plastic deformation at room temperature results in a change in the mechanical properties of metals: a general increase in ultimate tensile strength, yield point elongation, and hardness, while elongation and contraction decrease. However, the properties of metals can be reversed to their original level by heating them to a specific temperature for a specific time. Figure 1.15 schematically illustrates changes in the mechanical properties depending on the temperature to which a previously cold deformed workpiece of a steel of medium hardness has been raised.

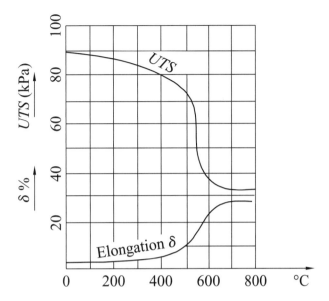

Fig. 1.15 Changes in the mechanical properties of steel of medium hardness during the process of recovery and recrystallization.

At the beginning of the reversal process, the material is slowly heated to the lower temperatures, then to a specific temperature for each material. At this point, sudden changes occur which correspond with the beginning of changes of the structurally deformed metal. In place of the deformed grains in a crystal lattice with defects, new crystallization centers appear, and around them new crystals are formed. Strains and defects of crystal lattice which appeared during plastic deformation now disappear.

This process is called *recrystallization*. The beginning of this process, i.e. the temperature at which new small grains with a crystal lattice form without defects, is called the temperature of recrystallization (T_r). This temperature is different for different metals and alloys, and may be calculated approximately by the following formula:

$$T_r = 0.4 \ T_m \tag{1.1}$$

where

T_m = the melting point of the metal.

Alloys usually have a higher temperature of recrystallization than pure metals. Changes during the process of heating before the temperature of recrystallization is reached are called *recovery*. The temperature for recovery (T_{rc}) ranges can be approximately calculated by the following formula:

$$T_{rc} = (0.2 \ to \ 0.3)T_m \tag{1.2}$$

During recovery, cold-work properties disappear, and there is no microscopic change, but there is residual stress.

Two

Mechanical Behavior of Materials

2.1 INTRODUCTION

One of the most important groups of processes in manufacturing is plastic deformation. The group includes forging, rolling, extrusion, rod and wire drawing, and all sheet-metal forming processes. This chapter discusses the fundamental aspects of the mechanical behavior of metals during deformation.

Metal deformation is an integral part of industrial production. For example, during the process of stretching a piece of metal to make some part of an aircraft or an automobile, the material is subjected to tension. By the same token, when a solid cylindrical piece of metal is forged in the making of a gear disk, the material is subjected to compression. Sheet metal, too, undergoes shearing stress when a hole is punched in it.

Strength, hardness, toughness, elasticity, plasticity, brittleness, ductility, and malleability are mechanical properties used as measurements of how metals behave under a load. These properties are described in terms of the types of force or stress that the metal must withstand and how these forces are resisted. Common types of loading are compression, tension, shear, torsion, or a combination of these stresses, such as fatigue. Figure 2.1 shows the three most common types of stress.

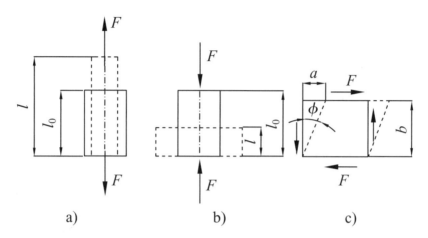

Fig. 2.1 Types of stress: a) tension, b) compression, c) shear.

Compression stresses develop within a sample material when forces compress or crush it. For example, the material that supports an overhead beam is in compression, and the internal stresses that develop within the supporting column are compressive.

Tension (or tensile) stresses develop when a material is subject to a pulling load — for example, when a wire rope is used to lift a load or as a guy to anchor an antenna. *Tensile strength* is defined as resistance to longitudinal stress or pull and can be measured in pounds per square inch of cross section:

Shearing stresses occur within a material when external forces are applied along parallel lines in opposite directions. Shearing forces can separate a material by sliding part of it in one direction and the remainder in the opposite direction.

Materials scientists learn about the mechanical properties of materials by testing them. Results from the tests depend on the size and shape of the material to be tested (the specimen), how it is held, and how the test is performed. To make the results comparable, use is made of common procedures, or *standards*, which are published by the ASTM.

To compare specimens of different sizes, the load is calculated per unit area. The force divided by the area is called stress. In tension and compression tests, the relevant area is that perpendicular to the force. In shear tests, the area of interest is perpendicular to the axis of rotation.

The engineering stress is defined as the ratio of the applied force F to the original cross/sectional area A_0 of the specimen. That is:

$$\sigma = \frac{F}{A_0} \qquad (2.1)$$

where

F = tensile or compressive force

A_0 = original cross-sectional area of the specimen

The nominal strain or engineering strain can be defined in three ways: As the relative elongation, given by:

$$\varepsilon_n = \frac{l - l_0}{l_0} = \frac{l}{l_0} - 1 \tag{2.2}$$

where

l_0 = the original gauge length

l = the instantaneous length

As the reduction of the cross-section area, given by:

$$\psi = \frac{A_0 - A}{A_0} = \frac{\Delta A}{A_0} = 1 - \frac{A}{A_0} \tag{2.3}$$

Or as the logarithmic strain, given by:

$$\phi = \ln \frac{A_0}{A} \tag{2.4}$$

where

A_0 = the original cross-section

A = the instantaneous cross-section area

These definitions of stress and strain allow test results for specimens of different cross-sectional area A_0 and of different length l_0 to be compared. It is generally accepted that tension is positive and compression is negative. Shear stress is defined as:

$$\tau = \frac{F}{A_0} \tag{2.5}$$

where

F = force is applied parallel to the upper and lower faces, each of which has an area A_0.

Shear strain is defined as:

$$\gamma = \frac{a}{b} = \text{tg}\Phi \tag{2.6}$$

Shear stresses produce strains according to:

$$\tau = G \cdot \gamma \tag{2.7}$$

where

G = shear modulus

Torsion is a variation of pure shear. Shear stress then is a function of applied torque shear strains related to the angle of twist.

Materials subject to uniaxial tension shrink in the lateral direction. The ratio of lateral strain to axial strain is called *Poisson's ratio*.

$$v = \frac{\varepsilon_x}{e_y} \tag{2.8}$$

where

ε_x = lateral strains

ε_y = axial strains

The theory of isotropic elasticity defines Poisson's ratio in the next relationship:

$$-1 < v \le 0.5 \tag{8a}$$

The Poisson ratio for most metals is between 0.25 and 0.35. Rubbery materials have Poisson's ratios very close to 0.5 and are therefore almost incompressible. Theoretical materials with a Poisson's ratio of exactly 0.5 are truly incompressible, because the sum of all their strains leads to a zero volume change. Cork, on the other hand, has a Poisson's ratio close to zero, which makes cork function well as a bottle stopper. The cork must be easily inserted and removed, yet it also must withstand the pressure from within the bottle.

The elastic modulus, shear modulus, and Poisson's ratio are related in the following way:

$$E = 2G(1 + v) \tag{2.9}$$

2.2 STRESS/STRAIN CURVES

The relationship between the stress and strain that a material displays is known as a stress/strain curve. It is unique for each material. Figure 2.2 shows the characteristic stress/strain curves for three common materials.

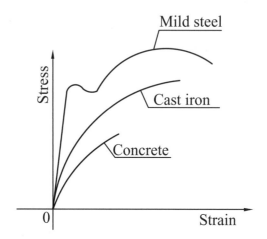

Fig.2.2 Stress/strain curves for different materials.

It can be seen that the concrete curve is almost a straight line. There is an abrupt end to the curve. This abrupt end, combined with the fact that the line is very steep, indicate that the material is brittle. The curve for cast iron is slightly curved. Cast iron also is a brittle material. Notice that the curve for mild steel seems to have a long gently curving "tail," which indicates the high ductility of the material. Typical results from a tension test on a mild-steel are shown in Fig. 2.3.

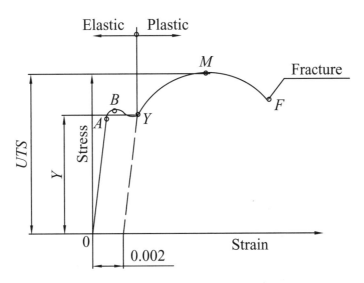

Fig. 2.3 Tensile stress/strain diagram for a mild steel.

Several significant points on a stress/strain curve help one understand and predict the way any building material will behave. Point *A* is known as the *proportional limit.* Up to this point, the relationship between stress and strain is exactly proportional. The linear relationship between stress and strain is called *Hooke's law.*

$$\sigma = E \cdot \varepsilon \tag{2.10}$$

If the load is removed, the specimen returns to its original length and shape, which is known as elastic behavior.

Strain increases faster than stress at all points on the curve beyond point *A*. Point *B* is known as the *elastic limit;* after this point, any continued stress results in permanent, or inelastic, deformation.

The stress resistance of the material decreases after the peak of the curve; this point is also known as the yield point *Y* of the material. Because the slope of the straight portion of the curve decreases slowly, the exact position on the stress /strain curve where yielding occurs may not be easily determined for soft and ductile materials. Therefore, *Y* is usually determined as the point on the stress/strain curve that is offset by a strain of 0.002 or 0.2% elongation.

If the specimen continues to elongate further under an increasing load beyond point *Y*, a domain curve begins in which the growth of strain is faster than that of stress. Plastic forming of metal is performed in this domain. If the specimen is released from stress between point *Y* and point *M,* the curve follows a straight line downward and parallel to the original slope (Fig. 2.4).

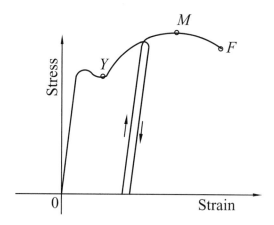

Fig. 2.4 Schematic illustration of loading and unloading tensile test specimen.

As the load and engineering stress increase further, the curve eventually reaches a maximum point *M*, and then begins to decrease. The maximum engineering stress in point *M* is called the ultimate tensile strength (UTS) of the material.

$$\sigma_m = UTS = \frac{F_{max}}{A_0} \qquad (2.11)$$

If the specimen is loaded beyond its ultimate tensile strength, it begins to "neck", or "neck down. The cross-sectional area of the specimen is no longer uniform along a gauge length, but is smaller in the necking region. As the test progresses, the engineering stress drops further and the specimen finally fractures at the point F. The engineering stress at fracture is known as the breaking or fracture stress.

The ratio of stress to strain in the elastic region is known as the *modulus of elasticity (E)* or Young's modulus and is expressed by:

$$E = \frac{\sigma}{\varepsilon} \qquad (2.12)$$

The modulus of elasticity is essentially a measure of the stiffness of the material.

2.3 DUCTILITY

Ductility is an important mechanical property because it is a measure of the degree of plastic deformation that can be sustained before fracture. Ductility may be expressed as either percent elongation or percent reduction in area. Elongation can be defined as:

$$\delta = \frac{l_f - l_0}{l_0} \times 100 \qquad (2.13)$$

Reduction can be defined as:

$$\psi = \frac{A_0 - A_f}{A_0} \times 100 \qquad (2.14)$$

where

l_f = length of the fracture. This length is measured between original gage marks after the pieces of the specimen are placed together

l_0 = the original sample gauge length

A_f = cross-sectional area at the fracture

A_0 = original sample gauge cross-sectional area

Knowledge of the ductility of a particular material is important because it specifies the degree of allowable deformation during forming operations. Gauge length is usually determined by inscribing gauge marks on the sample prior to testing and measuring the distance between them, before and after elongation has occurred. Because elongation is always declared as a percentage, the original gauge must be recorded. Fifty millimeters (two inches) is the standard gauge length for strip tensile specimens and this is how the data are generally recorded. The reduction in area area is declared as a percentage decrease in the original cross-sectional area and, like percent age elongation, it is measured after the sample fractures. The percentage elongation is more a measure of the strain leading to the onset of necking than a measure of the strain at final fracture in a uniaxial tensile specimen. A better measure of the strain at final fracture is the percentage reduction in area.

The relationship between the elongation and reduction of area is different for some groups of metals, as shown in Fig. 2.5.

Elongation ranges approximately between 10 and 60% for most materials, and values between 20 and 90% are typical for reduction of area. Thermoplastics and super-plastic materials, of course, exhibit much higher ductility, and brittle materials have little or no ductility.

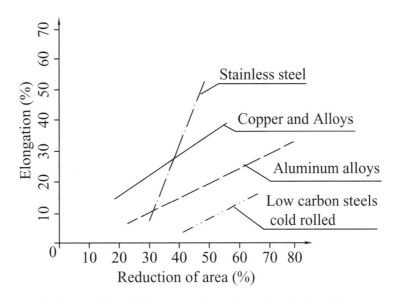

Fig. 2.5 Relationship between elongation and reduction of area.

2.4 TRUE STRESS AND TRUE STRAIN

In the solution of technical problems in the processes of sheet-metal forming, theoretical stress and strain do not have as crucial a significance as do true stress and true strain. True stress and true strain are much more important.

Because stress is defined as the ratio of force to area, It is apparent that true stress may be defined as:

$$k = \frac{F}{A} \tag{2.15}$$

where

A = the instantaneous cross-section area

As long as there is uniform elongation, true stress (k) can be expressed using the value for engineering stress. Assuming that volume at plastic deformation is constant: (this equation is in effect only to point M), the relationship between true and nominal stress may be defined as follows:

$$k = \frac{F}{A} = \sigma \frac{A_0}{A} = \sigma(1-\varepsilon) = \frac{\sigma}{1-\psi} = K\varepsilon^n \tag{2.15a}$$

where

K = is the strength coefficient

n = strain-hardening exponent

This equation is called the *flow curve*, and it represents the behavior of materials in the plastic zone, including their capacity for cold strain hardening.

Figure 2.6 shows a nominal (engineering) curve and the true stress and strain for medium carbon steel.

When the curve shown in Fig. 2.6a is plotted on a logarithmic graph, as in Fig. 2.6b, it is found that the curve is a straight line, and the slope of the line is equal to the exponent n known as *strain-hardening* (work-hardening) exponent. The strain-hardening exponent may have value from $n = 0$ (perfectly plastic solid) to $n = 1$ (elastic solid). For most metals, n has values between 0.10 and 0.50. The value of constant K equals the value of the true stress at a true strain value to 1.

Because the strains at the yield point Y are very small, the difference between the true and engineering yield stress is negligible for metals. This is because the difference in the cross- sectional areas A_0 and A at yielding is very small. However, the difference in the cross-sectional area A_0 and A, $(A < A_0)$ above point Y is always greater, so the difference between the true and nominal stress is significant $(k > \sigma)$. Because of the relationship, the given curve showing the true stress is known as the "hardening curve" of a metal.

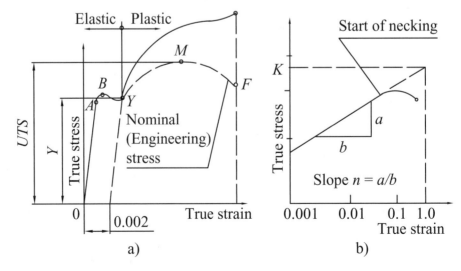

Fig. 2.6 Diagram of nominal and true stress: a) nonlogarithmic; b) logarithmic.

2.5 HARDNESS

Hardness is the property of a material that gives a general indication of its strength and of its resistance to plastic deformation, penetration, and wear. More specifically, the hardness of a material can be defined as its resistance to permanent indentation or abrasion. Therefore, hardness is important for many engineering applications, including most of the tooling used in manufacturing, because as hardness increases, so, generally, does resistance to wear by friction or erosion by steam, oil, or water.

2.5.1 Hardness Tests

Several methods have been developed for testing the hardness of materials. However, only three types of tests are generally used by the metals industry:

- Brinell hardness test
- Rockwell hardness test
- Vickers hardness test

a) Brinell Hardness Test

The Brinell hardness test is one of the earliest standardized methods for measuring the hardness of low-to-medium hard metals. It is determined by forcing a hard steel or cemented carbide ball of 10 mm in diameter under varying loads: 500 kg for soft metals such as copper, brass, and thin sheets; 1500 kg for aluminum castings; and 3000 kg for metals such as iron and steel. The load is usually applied for 15 to 30 seconds. The load and ball are removed and the diameter of the resulting spherical indentation is measured to ± 0.05 mm using a low-magnification portable microscope. The Brinell hardness number is calculated using the following equation:

$$HB = \frac{2F}{\left(\pi D_b\right)\left(D_b - \sqrt{D_b^2 - D_i^2}\right)} \tag{2.16}$$

where

HB = Brinell hardness number

D_b = diameter of the ball, mm

D_i = diameter of the indentation, mm

F = indentation load, kg

These dimensions are shown in Fig. 2.7. The resulting Brinell hardness number has units of kg/mm^2, but these are usually omitted in expressing the number. A well-expressed Brinell hardness number in technical documentation looks like this: 80 HB 10/500/15, which means that the Brinell hardness of 80 was obtained using a 10 mm diameter hardened steel ball with a 500 kg load for a period of 15 seconds. For harder materials above 500 HB, a hardened steel ball is used cemented carbide ball instead.

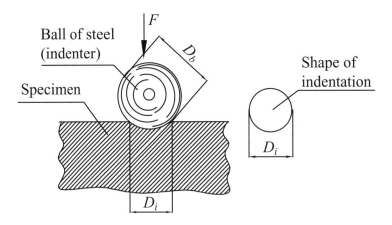

Fig. 2.7 Brinell hardness testing method.

b) Rockwell Hardness Test

The Rockwell hardness test is similar to the Brinell test in that the hardness value is a function of the indentation of a test piece by an indenter under a static load. In the Rockwell test, the indenter is either a diamond cone with an included angle of 120° and 0.2 mm tip radius, or a steel ball with a diameter of 1/16 in. Hardness numbers have no units and are commonly given in the A, B, or C scales. The higher the number is in each of the scales, the harder the material. It is determined by applying a preload first to the indenter using a load of 10 kg to seat it in the material and the indicator is then set at zero. The major load corresponding to a total of (60 kg A scale; 100 kg B scale; 150 kg C scale) is then applied and the indentation dent e is measured after the load is removed. Upon removal of the major load, the depth e is converted into a Rockwell hardness reading by the testing machine while the minor load is still on. A schematic of the Rockwell testing method is illustrated in Fig. 2.8.

The standard Rockwell test machine should not be used on materials less than 1.6 mm (1/16 in.) thick, on rough surfaces, or on materials that are not homogeneous, such as gray cast iron. Because of the small size of the indentation, localized variations of roughness, composition, or structure can greatly influence the results.

In contrast to the Brinell test, the Rockwell test offers the benefit of direct readings in a single step. The Rockwell test delivers rapid and reliable results, especially for routine inspections.

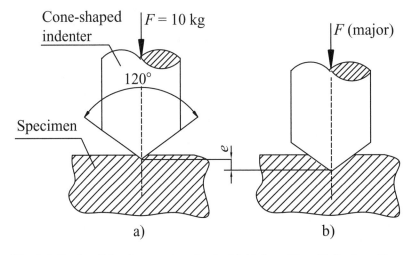

Fig. 2.8 Rockwell hardness test method: a) initial position; b) final position.

c) Vickers Hardness Test

The Vickers hardness test was developed in 1922 year. It uses a diamond pyramid with an indenter angle of 136° and a square base. As in the Brinell test, the Vickers hardness number is the ratio of the load to the surface area of the indentation in kilograms per square millimeter. The diamond is pressed into the surface of the material at loads ranging between 2 and 120 kg in standardized steps, depending on the material.

To perform the Vickers test, the specimen is placed on an anvil that has a screw-threaded base. The anvil is turned, raising it by the screw threads until it is close to the point of the indenter. With the start lever activated, the load is slowly applied to the indenter. The load is released and the anvil with the specimen is lowered. The operation of applying and removing the load is controlled automatically. A schematic illustration of the Vickers test is shown in Fig. 2.9.

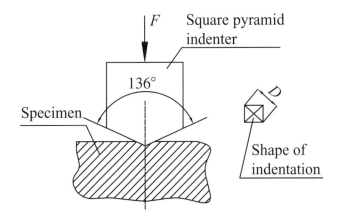

Fig. 2.9 Vickers hardness test method.

Several loadings give practically identical hardness numbers on uniform materials, which is much better than the arbitrary changing of scale with the other hardness machines. A filar microscope is swung over the specimen to measure the square indentation to a tolerance of ± 0.001 mm. The advantages of the Vickers hardness test are these:

- Extremely accurate readings can be taken.
- Just one type of indenter is used for all types of metals and surface treatments.
- It is simple to use.
- Little time is involved.
- Little surface preparation is required.
- Testing can be done on location.
- The test is relatively inexpensive.

The Vickers number is calculated using the following formula:

$$HV = \frac{1.854F}{D^2}$$ (2.17)

where

HV = Vickers number

F = applied load, kg

D = diagonal of the impression, mm

d) Other Hardness Test Methods

Other common hardness test methods include the following: scleroscope, durometer, microhardness test, and Mohs hardness test.

Scleroscope. The scleroscope is a portable instrument that measures the rebound height of a diamond-tipped indenter (hammer) dropped from a fixed height onto the surface of a material being tested. It is simply placed on the surface of the part, and it is particularly useful in limited spaces where conventional hardness test device (testers) cannot be used. The scleroscope measures the mechanical energy absorbed by the material when the indenter strikes the surface on which the test is made. This surface must be fairly highly polished to obtain a good result. The hardness is related to the rebound of the indenter; the higher the rebound, the harder the material. Primarily used to measure the hardness of large parts of steel and other ferrous metals, scleroscope hardness numbers are comparable only among similar materials.

Durometer. The durometer is used for the hardness testing of very soft elastic materials such as rubbers and plastics. In the test, the resistance to elastic deformation is expressed as a hardness number. An indenter is pressed against the surface, and then a constant load is applied rapidly. The depth of penetration is measured after one second.

Microhardness test. One of the most used microhardness tests is the Knoop test, which is used when it is necessary to determine hardness over a very small area of a material. The position for the test is selected under high magnification. In this test, a pyramid-shaped diamond indenter with apical angles of 130° and 170° is loaded with a predetermined load of 25 to 3600 grams against a material. The diamond indenter leaves a four-sided impression with one diagonal seven times longer than the other (ratio 7:1). The length of the impressions is determined using a microscope, because the mark is very small. The hardness number is determined by the following equation:

$$HK = \frac{14.23F}{D^2} \tag{2.18}$$

where

HK = Knoop hardness number

F = applied load, kg

D = longer diagonal, mm

Mohs hardness test. This test, also known as the *scratch hardness test*, can be defined as the ability of the material to resist being scratched. To give numerical values to these properties, the German mineralogist Friedrich Mohs defined the *Mohs hardness scale*, in which 10 minerals are ranged in order from 1 to 10, with 1 being the measure for talc and 10 for diamond (the hardest substance known). According to this scale, a given material with a higher Mohs hardness number always scratches one with a lower number. In this manner, any substance can be assigned an approximate number on the Mohs scale. For example, ordinary glass would be 5.5, hardened steel about 6, and aluminum oxide for cutting tools about 9, etc. This test is not used for manufacturing processes, but it is quite useful in mineral identification.

2.6 EFFECT OF TEMPERATURE ON MATERIAL PROPERTIES

Metals are generally specified for operating at or about the ambient (room) temperature. When a metal is used in a range outside this temperature, its properties are significantly affected. It cannot be overemphasized that test data used in design and engineering decisions should be obtained under conditions that best simulate the conditions of service. Consequently, it is very important for the designers of structures such as aircraft, space vehicles, gas turbines, as well as chemical processing equipment that require operation under low or high temperatures, to know both the short-range and long-range effects of temperature on the mechanical and physical properties of the materials being used for such applications.

From the manufacturing standpoint, the effects of temperature on material properties are also important because numerous manufacturing processes depend on the properties of a material, which may be affected by its temperature in a favorable or unfavorable manner. In general, an increase in temperature tends to promote a drop in strength and hardness properties, but also an increase in elongation. For sheet metal forming processes, these trends are of considerable importance because they permit forming to be done more readily at elevated temperatures at which a material is weaker and more ductile.

Strength and elongation. Most ferrous metals have a maximum strength at approximately 200°C. The strength of non-ferrous metals is generally at a maximum at room temperature.

Table 2.1 gives the strength of some ferrous metals as a percentage of their strength at 20°C when the operating temperature is increased.

Table 2.1 The strength of metals at operating temperature as a percentage of their strength at 20°C

Material	Temperature °C				
	100	200	300	400	500
Wrought Iron	104	112	111	96	76
Cast Iron	–	100	99	92	76
Steel Castings	109	125	115	97	57
Structural Steel	103	132	122	86	49
Copper	95	83	73	59	42
Bronze	101	94	57	26	

At temperatures below –10°C, typical steels become more brittle and their toughness is affected. With selected grades of carbon steel, it is possible to operate at temperatures as low as –40°C. At these low temperatures, it is necessary to conduct specific impact tests at low temperatures on the selected steel to confirm suitability. Austenitic stainless steel is particularly useful in cryogenic applications because of its high toughness and strength at very low temperatures. 304 stainless steel can be readily employed to handle liquid helium and liquid hydrogen (–268,9 and –252.7°C respectively). Figure 2.10 shows the general effect of temperature on strength and elongation.

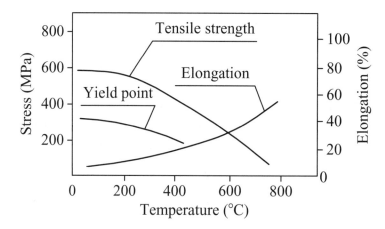

Fig. 2.10 Some effects of temperature on mechanical properties of steel:
a) tensile strength; b) yield strength; c) elongation.

Young's modulus. The variation of Young's modulus for ferrous and most engineering materials needs to be considered only up to 1,000°C, because at greater temperatures these materials have no mechanical resistance. Figure 2.11 shows the effect of temperature on Young's modulus, where $E_{(T)}$ is Young's modulus at operating temperature, $E_{(20)}$ is Young's modulus at 20°C, and T°C is the operating temperature.

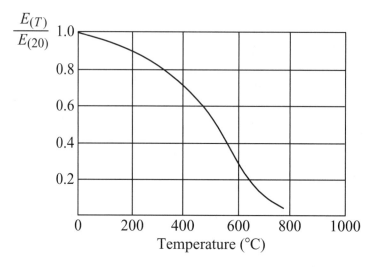

Fig. 2.11 Effect of temperature on Young's modulus.

Poisson's ratio. Materials subjected to uniaxial tension will shrink in the lateral dimension. The ratio of lateral strain to axial strain is called Poisson's ratio. Poisson's ratio for most metals is between 0.25 and 0.35 at room temperature. Figure 2.12 summarizes the data for Poisson's ratio v as a function of temperature.

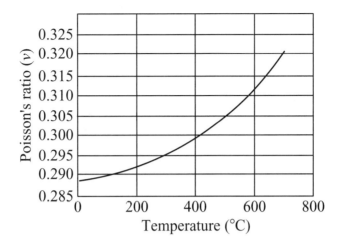

Fig. 2.12 Poisson's ratio as a function of temperature.

Impact. The effect of temperature on impact properties becomes the subject of intense study when a system is located in a cold environment. Figure 2.13 shows impact properties as a function of decreasing temperature on carbon steel. The temperature at which the response goes from high to low energy absorption is known as the *transition temperature*; it can be different for similar compositions of steel, but all steels tend to exhibit the rapid transition in impact stretch when the temperature decreases. Knowing the transition temperature is useful in evaluating the suitability of materials for certain applications.

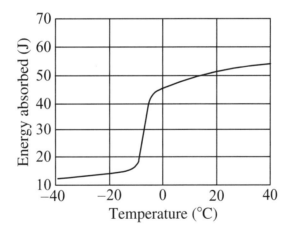

Fig. 2.13 Illustration of impact properties as a function of temperature.

Creep. At elevated temperatures and constant stress or load, many materials continue to deform at a slow rate with time. This behavior is called creep. Creep in steel is important only at elevated temperatures. In general, creep becomes significant at temperatures above $0.4T_m$ where T_m is the absolute melting temperature. However, materials having low melting temperatures will exhibit creep at ambient temperatures. Good examples are lead and various types of plastic. For example, lead has a melting temperature of 326°C, and at 20°C it exhibits similar creep characteristics to those of iron at 650°C. Although the rate of elongation is small, it is very important in the design of such equipment as steam or gas turbines, as well as high-temperature pressure vessels that operate at high temperatures for long periods of time. Figure 2.14 shows strain as a function of time due to constant stress over an extended period at a fixed temperature.

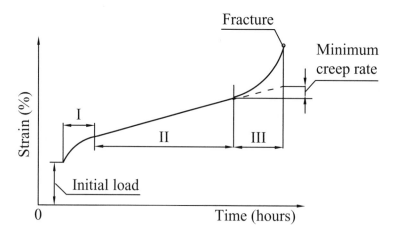

Fig. 2.14 Creep as a function of time at a fixed elevated temperature.

The curve contains three distinct stages: I (primary), II (secondary), and III (tertiary). Initial or primary creep occurs in Stage I. The creep rate is relatively high at first, but it soon slows with increasing time. In Stage II, the creep rate is minimum, and strain increases very slowly with time. In Stage III, the creep rate exponentially increases, and strain may become so large that it leads to failure.

Part
Two

Manufacturing Processes

This part discusses the methods that are most used for sheet metal forming. The text is presented in five chapters, which deal in detail with: shearing; punching and blanking; bending; deep drawing; and various forming processes (stretching, nosing, expanding, flanging, flexible die forming, and spinning). Special attention is given to the mechanism of processes, estimation of forces, clearances, minimum bend radius, and other important factors in sheet-metal forming processes. Detailed analytical mathematical transformations are not included, but the final formulas derived from them, which are necessary in practical applications, are included.

Three

Shearing Process

3.1 MECHANICS OF SHEARING

The shearing process involves the cutting of flat material forms, such as sheets and plates. The cutting may be done by different types of blades or cutters in special machines driven by mechanical, hydraulic, or pneumatic power. Generally the operations consist of holding the stock rigidly while it is severed by the force of an upper blade as it moves down past the stationary lower blade.

During the shearing process, three phases may be noted. In phase I, because of the action of the cutting force F, the stress on the material is lower than the yield stress ($\tau < \tau_e$). This phase is that of elastic deformation (Fig. 3.1). To prevent the movement of material during the cutting operation, the material is held by the material holder at force F_d. In phase II, the stress on the material is higher than the yield stress but lower than the *UTS*. This phase is that of plastic deformation ($\tau_e < \tau < \tau_m$). In phase III, the stress on the material is equal to the shearing stress ($\tau = \tau_m$). The material begins to part not at the leading edge, but at the appearance of the first crack or breakage in the material. Fracture of the material occurs in this phase.

The amount of penetration of the upper blade into the material is governed by the ductility and thickness of the work material. If the material is thicker and more brittle, the first crack will appear earlier, so there is earlier disjunction of the material. The sheared edge is relatively smooth where the blade penetrates, with a considerably rougher texture along the torn portion.

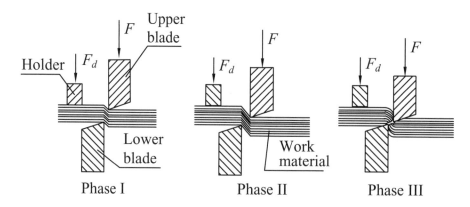

Fig. 3.1 Schematic illustration of the shearing process.

3.2 SHEARING FORCES

Knowledge of the forces and the power involved in shearing operations is important, and they may be calculated according to the edge types of the cutters. There are three types of cutters:

- straight parallel cutters
- straight inclined cutters
- rotary cutters.

3.2.1 Shearing with Straight Parallel Cutters

The shearing force F with straight parallel cutters (Fig 3.2) can be calculated approximately as:

$$F = \tau \cdot A \tag{3.1}$$

where

  $A =$ cutting area

  $\tau =$ shear stress

The cutting area is calculated as:

$$A = b \cdot T \tag{3.1a}$$

where

b = width of material

T = thickness of material

This calculating force needs to be increased by 20% to 40% depending on the:

- obtuseness of the cutter edge
- enlarged clearance between cutters
- variations in the thickness of the material
- other unpredictable factors

The real force of the shearing machine is:

$$F_M = 1.3F \qquad (3.2)$$

The crosscut force F_t at the cutters (Fig. 3.2) is:

$$F_t = F \cdot tg\gamma \qquad (3.3)$$

For shearing without a material holder, the turn angle of the material is:

$$\gamma = 10^0 \text{ to } 20^0$$

and the crosscut force is:

$$F_t = (0.18 \div 0.36)F \qquad (3.3a)$$

For shearing with a material holder the turn angle is:

$$\lambda = 5^0 \text{ to } 10^0$$

and the crosscut force

$$F_t = (0.09 \div 0.18)F \cdot b \qquad (3.3b)$$

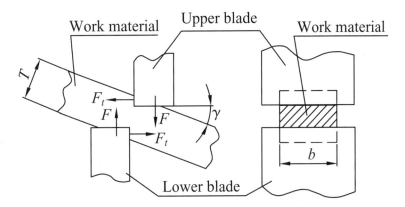

Fig. 3.2 Schematic illustration of shearing with straight parallel cutters.

3.2.2 Shearing with Straight Inclined Cutters

Shears with straight inclined cutters (Fig. 3.3) are used for cutting material of relatively small thickness compared with the width of cutting. Using inclined cutters reduces the shearing force and increases the range of movement necessary to disjoin the material. The penetration of the upper blade into material is gradual and, as result, there is a lower shearing force.

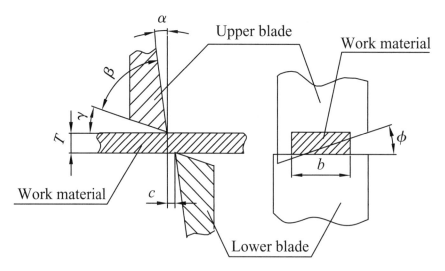

Fig. 3.3 Schematic illustration of shearing processes with straight inclined cutters.

The shearing force can be calculated approximately as:

$$F = n \cdot k \cdot UTS \cdot \varepsilon_{ot} \frac{T^2}{\text{tg}\phi} \tag{3.4}$$

where

n = 0.75 to 0.85 (for most materials)
k = 0.7 to 0.8 (ratio UTS/s for material)
e_{ot} = the relative amount of penetration of the upper blade into the material (Table 3.1)
ϕ = angle of inclination of the upper cutter

Table 3.1 Relative amount of penetration of the upper blade into the material

MATERIAL	Thickness of material T (mm)			
	< 1	1 to 2	2 to 4	>4
Plain carbon steel	0.75 – 0.70	0.70 – 0.65	0.65 – 0.55	0.50 – 0.40
Medium steel	0.65 – 060	0.60 – 0.55	0.55 – 0.48	0.45 – 0.35
Hard steel	0.50 – 0.47	0.47 – 0.45	0.44 – 0.38	0.35 – 0.25
Aluminum and copper (annealed)	0.80 – 0.75	0.75 – 0.70	0.70 – 0.60	0.65 – 0.50

For shearing without a material holder, the angle of inclination of the cutter is $\gamma = 7^0$ to 12^0. If the angle is $\gamma > 12^0$, the shears must have a material holder. To make the shearing process efficient, the cutters are made (see Fig. 3.3) with:

- end relief angle $\gamma = 3^0$ to 12^0
- back angle $\alpha = 2^0$ to 3^0
- lip angle $\beta = 75^0$ to 85^0

The clearance between cutters is c = (0.02 to 0.05) mm.

3.2.3 Shearing with Rotary Cutters

The rotary shearing process is very much like shearing with straight inclined cutters because the straight blade may be thought of as a rotary cutter with an endless radius. It is possible to make straight line cuts as well as to produce circular blanks and irregular shapes by this method. Figure 3.4 illustrates the conventional arrangement of the cutters in a rotary shearing machine for the production of a perpendicular edge. Only the upper cutter is rotated by the power drive system. The upper cutter pinches the material and causes it to rotate between the two cutters.

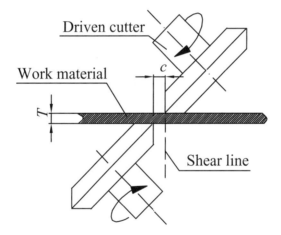

Fig. 3.4 Schematic illustration of shearing with rotary cutters.

Shearing force can be calculated approximately as:

$$F = 0.6 \cdot UTS \cdot \varepsilon_{ot} \frac{T^2}{2 tg\phi} \tag{3.5}$$

The clearance between rotary cutters with parallel inclined axes is:

$$c = (0.1 \text{ to } 0.2)T \tag{3.6}$$

Rotary shearing machines are equipped with special holding fixtures that rotate the work material to generate the desired circle. A straight edge fixture is used for straight line cutting. It should be noted that the rotary shearing operation, if not performed properly, may cause various distortions of the sheared part.

Four

Blanking and Punching

4.1 BLANKING AND PUNCHING MECHANISM

Blanking and punching are processes used to cut metal materials into any precise form by the use of dies. The basic parts of the tool are the punch and the die. The major variables in these processes are as follows: the punch force, F; the speed of the punch; the surface condition and materials of the punch and die; the condition of the blade edges of the punch and die; the type of lubricant; and the amount of clearance between the punch and die.

In blanking, the slug is the part that is used, and the remainder is scrap. In punching, the sheared slug is discarded as scrap, leaving the remainder to be used (Fig. 4.1).

43

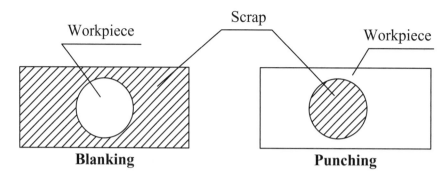

Fig. 4.1 Blanking and punching.

There are three phases in the process of blanking and punching (Fig. 4.2.). In phase I, during which the work material is compressed across and slightly deformed between the punch and die, the stress and deformation in the material do not exceed the elastic limit. This phase is known as the *elastic phase*.

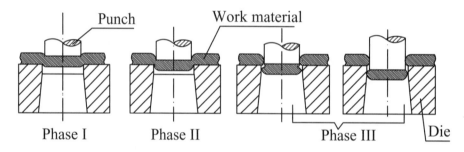

Fig. 4.2 Phases in the process of shearing.

In Phase II, the bent work material is pushed farther into the opening die by the punch. At this point in the operation, the material has been obviously deformed at the rim, between the cutting edges of the punch and die. This concentration of outside forces causes plastic deformation at the rim of the material. At the end of this phase, the stress in the work material close to the cutting edges reaches a value corresponding to the material shear strength, but the material resists fracture. This phase is called the plastic phase.

During Phase III, the strain in the work material reaches the fracture limit. Micro-cracks appear which turn into macro-cracks, followed by separation of the parts of the workpiece. The cracks in the material start at the cutting edge of the punch on the upper side of the work material, also at the die edge on the lower side of the material; the cracks propagate along the slip planes until complete separation of the part from the sheet occurs. A slight burr is generally left at the bottom of the hole and at the top of the slug. The slug is then pushed farther into the die opening. The slug burnish zone expands and is held in the die opening. The whole burnish zone contracts and clings to the punch. The overall features of the blanked or punched edges for the entire sheared surface are shown in Fig. 4.3. Note that the edges are neither smooth nor perpendicular to the plane of the sheet.

The clearance *c* is the major factor determining the shape and quality of the blanked or punched edge. If clearance increases, the edges become rougher, and the zone of deformation becomes larger. The material is pulled into the clearance area, and the edges of the punched or blanked zone become more and more rounded. In fact, if the clearance is too large, the sheet metal is bent and subjected to tensile stresses instead of undergoing a shearing deformation.

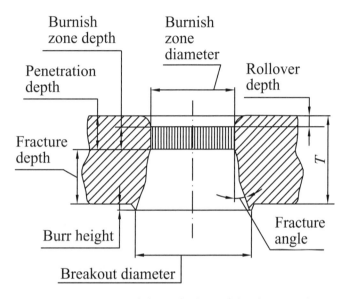

Fig. 4.3 Features of sheared edges of the sheet metal.

4.2 BLANKING AND PUNCHING CLEARANCE

Clearance *c*, is the space (per side) between the punch and the die opening (Fig. 4.4), such that:

$$c = \frac{D_{\mathrm{m}} - d_{\mathrm{p}}}{2} \tag{4.1}$$

Because of the amount of clearance between the punch and the die, tool and die producers enjoy some kind of mystique related to their work as being both an art and a science.

Proper clearance between cutting edges enables the fractures to start ideally at the cutting edge of the punch and also at the die. The fractures will proceed toward each other until they meet, and the fractured portion of the sheared edge then has a clean appearance. For optimum finish of a cut edge, correct clearance is necessary and is a function of the kind, thickness, and temper of the material.

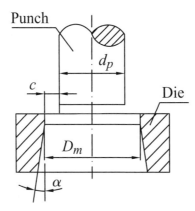

Fig. 4.4 Punch and die clearance.

The upper corner of the cut edge of the strip stock and the lower corner of the blank acquire a radius where the punch and die edges respectively make contact with the work material. This edge radius is produced by the plastic deformation taking place and is more pronounced when cutting soft materials. Excessive clearance will also cause a large radius at these corners, as well as a burr on opposite corners.

When clearance is not sufficient, additional layers of the material must be cut before complete separation is accomplished. With correct clearance, the angle of the fractures will permit a clean break below the burnish zone because the upper and lower fractures will extend toward one another. Excessive clearance will result in a tapered cut edge, because for any cutting operation, the opposite side of the material that the punch enters after cutting will be the same size as the die opening.

The width of the burnish zone is an indication of the hardness of the material. Provided that the die clearance and material thickness are constant, the softer the material is, the wider will be the burnish zone. Harder metals required larger clearance and permit less penetration by the punch than ductile materials; dull tools (punch and die) create the effect of too small a clearance as well as a burr on the die side of the stock. Clearance is generally expressed as a percentage of the material thickness, but some authorities recommend absolute values.

Table 4.1 illustrates the value of the shear clearance in percentages depending on the type and thicknesses of the material. Table 4.2 illustrates absolute values for the shear clearance depending on the type and thickness of the material.

Table 4.1 Values for double clearance ($2c$) as a percentage of the thickness of the materials

MATERIAL	Material thickness T (mm)				
	< 1.0	1.0 to 2.0	2.1 to 3.0	3.1 to 5.0	5.1 to 7.0
Low carbon steel	5.0	6.0	7.0	8.0	9.0
Copper and soft brass	5.0	6.0	7.0	8.0	9.0
Medium carbon steel 0.20% to 0.25% carbon	6.0	7.0	8.0	9.0	10.0
Hard brass	6.0	7.0	8.0	9.0	10.0
Hard steel 0.40% to 0.60% carbon	7.0	8.0	9.0	10.0	12.0

Table 4.2 Absolute values for double clearance (2*c*) for some stock materials

Material thickness T (mm)	Material			
	Low carbon steel, copper and brass	Medium steel 0.20% to 0.25% carbon	Hard steel 0.40% to 0.60% carbon	Aluminum
0.25	0.01	0.015	0.02	0.01
0.50	0.025	0.03	0.035	0.05
1.00	0.05	0.06	0.07	0.10
1.50	0.075	0.09	0.10	0. 15
2.00	0.10	0.12	0.14	0.20
2.50	0.13	0.15	0.18	0.25
3.00	0.15	0.18	0.21	0.28
3.50	0.18	0.21	0.25	0.35
4.00	0.20	0.24	0.28	0.40
4.50	0.23	0.27	0.32	0.45
4.80	0.24	0.29	0.34	0.48
5.00	0.25	0.30	0.36	0.50

More analytical methods for the definition of shear clearance exist; for instance, the theoretical method expresses clearance as a function of thickness of the material and the shear strength of the material. That is:

$$c = \frac{k \cdot T \sqrt{\tau_m}}{6.32} = \frac{k \cdot T \sqrt{0.7UTS}}{6.32} \quad \text{for } T \leq 3 \text{ mm} \tag{4.2}$$

$$c = \frac{(k \cdot T - 0.015)\sqrt{0.7UTS}}{6.32} \quad \text{for } T \leq 3 \text{ mm}$$

where

T = thickness of material

k = coefficient that depends on the type of die; that is $k = 0.005$ to 0.035, most frequently uses $k = 0.01$. For a die of metal ceramic, $k = 0.015$ to 0.18

τ_m = shear strength of material

UTS = Ultimate Tensile Strength

Location of the proper clearance determines either hole or blank size; the punch size controls the hole size, and the die size controls the blank size. A blank will be sheared to size when the punch is made to the desired size less the amount of clearance. A punched hole of the desired size results when the die is made to the desired size plus the amount of clearance.

4.3 PUNCH FORCE

Theoretically, punch force should be defined on the basis of both tangential (τ) and normal (σ) stresses that exist in a shear plan. However, with this analysis, one obtains a very complicated formula that is not convenient for use in engineering practice because an estimate of the punching force can be calculated from a value for tangential stress alone.

4.3.1 Punch and Die with Parallel-Cut Edges

The force F for a punch and die with parallel-cut edges is estimated from the following equation:

$$F = LT\tau_m = 0.7LT(UTS) \tag{4.3}$$

where

    L = the total length sheared (perimeter of hole)

    T = thickness of the material

    UTS = the ultimate tensile strength of the material

Such variables as unequal thickness of the material, friction between the punch and the workpiece, or poorly sharpened edges, can increase the necessary force by up to 30%, so these variables must be considered in selecting the power requirements of the press. That is, the force requirements of the press is:

$$F_p = 1.3F \tag{4.4}$$

The shared zone is subjected to cracks, plastic deformation ,and friction, all of which affect the punch force/penetration curves, which can thus have various shapes. One typical curve for a ductile material is shown in Fig. 4.5. The area under this curve is the total work done in the punching or blanking operation.

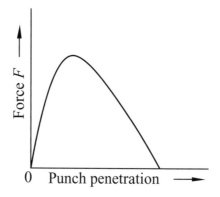

Fig. 4.5 Typical punch force/penetration curve in shearing.

4.3.2 Punch and Die with Bevel-Cut Edges

The punch force can be reduced if the punch or die has bevel-cut edges (Fig. 4.6). In blanking operations, shear angles, which may be convex or concave bevel shear, should be used (Fig. 4.6a) to ensure that the slug workpiece remains flat. In punching operations, bevel shear or double-bevel shear angles should be used on the punch (Fig. 4.6b). The height of the bevel shear and the shear angle depend on the thickness of material and they are:

$$H \leq 2T \text{ and } \varphi \leq 5^{0} \text{ for } T \leq 3 \text{ mm}$$

$$H = T \text{ and } \varphi \leq 8^{0} \text{ for } T > 3 \text{ mm} \tag{4.5}$$

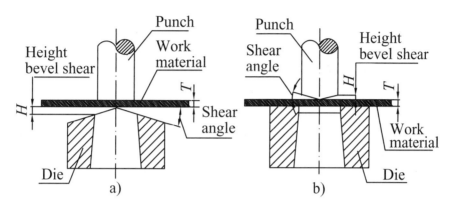

Fig. 4.6 Shear angles: a) on die; b) on punch.

Increases in the shear angle also increase deformations, which can cause warping of the workpiece. The punch force F when a shear angle is used on a punch and die in shearing operations is estimated in the following equation:

$$F = 0.7k \cdot L \cdot T(UTS) \tag{4.6}$$

where

$$k = 0.4 \text{ to } 0.6 \text{ for } H = T$$

$$k = 0.2 \text{ to } 0.4 \text{ for } H = 2T$$

Use of a shear angle on the punch or die for punching and blanking workpieces with complicated shapes is not recommended.

4.4 MATERIAL ECONOMY

The major portion of the cost of producing a stamped component is the material. Material economy is therefore of the utmost importance from the standpoint of cost. The relative position of the blanks on the work material should be carefully laid out to avoid unnecessary scrap. For the greatest material economy it is necessary to plan:

- A rational scrap strip layout
- An optimum value of m (from the edge of the blank to the side of the strip) and n (the distance from blank to blank)
- Techniques adapted for altering the layout without changing the function of the workpiece
- Utilization of the scrap from one part as material for another

The best criterion for estimating material economy is assessing the percentage of scrap. A good rule to follow is to lay out blanks in such a way as to utilize at least 70 to 80% of the strip or sheet area.

4.4.1 Scrap Strip Layouts

In general, there are three types of scrap strip layout. The first uses the value of *n* and *m* (Fig. 4.7a). The second uses only the value of *n*, but *m* = 0 (Fig. 4.7b). The third uses the value of *n* = 0 and *m* = 0 (Fig.4.7c). The scrap strip layout is a good place to start when designing a die. Strips of stock are fed into the die from the right, from the left, or from front to rear, depending upon the method selected. Strips may be sheared to desired widths at the mill, or they may be purchased as rolled stock.

Assume a straightforward single-pass layout, as shown in Fig. 4.7a. Before the material requirements can be calculated, the value of *m* and *n* must be determined.

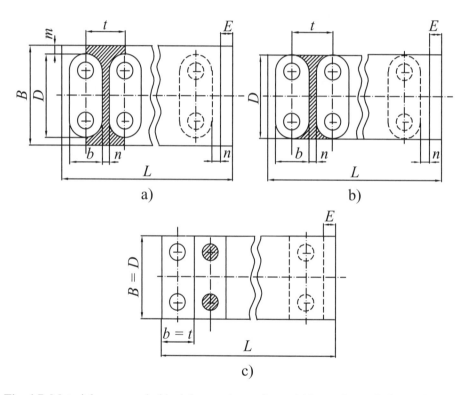

Fig. 4.7 Material economy in blank layout a) $m > 0$; $n > 0$ b) $m = 0$; $n > 0$ c) $m = 0$; $n = 0$.

If the thickness of the materials T = 0.6 mm, the values of *m* and *n* may be taken from Table 4.3. When the thickness of the material T > 0.6 mm, the value of *m* and *n* may be calculated from the formula

$$m = T + 0.015D \qquad (4.7)$$

where

   m = distance from the edge of the blank to the side of the strip

   T = thickness of the material

   D = width of blank

Table 4.3 Absolute values for *m* and *n*

THICKNESS OF THE MATERIAL $T \le 0.6$ (mm)				
DIMENSION (mm)	**Strip width B (mm)**			
	Up to 75	**76 to 100**	**101 to 150**	**151 to 300**
m & n	2.0	3.0	3.5	4.0
DIMENSION (mm)	**THICKNESS OF THE MATERIAL $T > 0.6$ (mm)**			
	0.61 to 0.8	**0.81 to 1.25**	**1.26 to 2.5**	**2.51 to 4.0**	**4.1 to 6.0**
n	3.5	4.3	5.5	6.0	7.0

The number of blanks *N* which can be produced from one length of stock is:

$$N = \frac{L - n}{t} \qquad (4.8)$$

where

    L = length of stock

    t = length of one piece

    N = number of blanks

The value of *n* is found from Table 4.3.

The scrap at the end of the strip *E* is:

$$E = L - \left(N \cdot t + n\right) \qquad (4.9)$$

The position of the workpiece relative to the direction of feed of the strip is an important consideration. Generally speaking, there are two types: rectangular (one axis of the blank is parallel with stock movement through the die) and at an angle (no axis of the blank is parallel to the direction of feed).

Figure 4.8 shows four possible layouts of L-shaped parts. The rectangular layout, Fig. 4.8a, utilizes 62.5% of the strip. The angle layout with adequate *n* dimension, Fig. 4.8b, utilizes 76.5% of the strip. The layout in Fig. 4.8c utilizes 81.8%. The greatest material economy is achieved with the layout type in Fig. 4.8d, but design and production of the die are more complicated than in the other layouts shown.

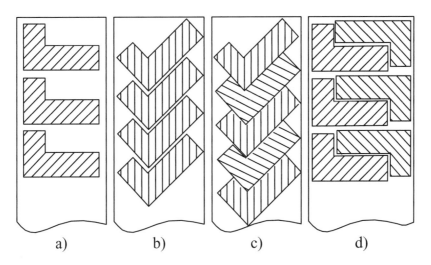

      a)   b)   c)   d)

Fig. 4.8 Material economy in blank layout of L-shaped parts.

4.4.2 Altering the Design of the Workpiece

A savings in stock may be obtained by using adaptive techniques to alter the layout of the workpiece. Figure 4.9 illustrates how modifying the contour of a commercial stamping can permit nesting and reduce 40% scrap loss while still retaining the functionality of the product

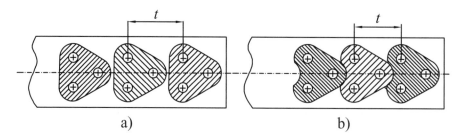

Fig. 4.9 Material economy by adapting the technique in the altered layout: a) original layout; b) altered layout.

4.4.3 Multi-Line Layout

Figure 4.10 illustrates an alternate multi-line layout of circular blanks. This type of layout improves material economy and is used for smaller, simple-shaped workpieces. To simplify the die, multiple sets of punches may stamp out blanks simultaneously. The strip width B may be calculated from the following formula:

$$B = D + 0.87(D + n)(i + 1) + 2m \qquad (4.10)$$

where

D = width of blank

m = edge of blank to side of strip

n = distance of blank to blank

i = number of lines

The value of i may be taken from Table 4.4.

Table 4.4 Number of lines i in an alternate multi-line layout

Blank width D (mm)	< 10	10 to 30	30 to 75	75 to 100	> 100
Number of lines i	10	9 to 5	4 to 3	3 to 2	2 to 1

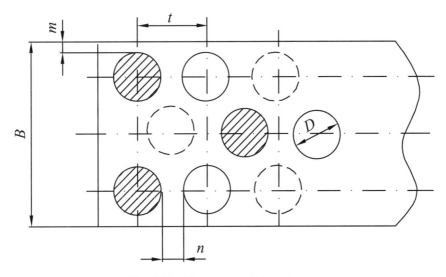

Fig. 4.10 Alternate multi-line layout.

4.4.4 Utilizing Scrap from One Piece as Material for Another Piece

Rings or squares with large holes are very wasteful of material. Sheet-metal stamping companies are constantly on the alert to utilize the scrap from one part as material for another. For example, in Fig. 4.11, the scrap center that otherwise would result in waste from piece number one is used as material for piece number two, and waste from piece number two is used as material for piece number three.

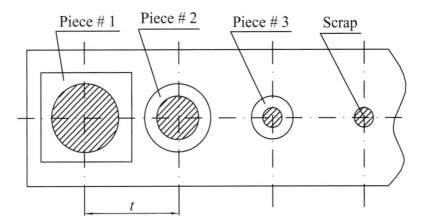

Fig. 4.11 Utilize the scrap from one piece as a material for another piece (scrap from piece #1 as the material for piece #2; scrap from piece #2 as the material for piece #3.

4.5 SHAVING

Sometimes it is necessary to achieve a very clean hole. In order to do this, a process called shaving is used. This is the process of the removal of a thin layer of material with a sharp punch. It can be used with a blanked and punched workpiece.

4.5.1 Shaving a Punched Workpiece

The edges of punched pieces are generally unsquare, rough, and uneven. This process is particularly useful in:

- improving the surface finish of a punched hole
- improving the dimensional accuracy of punched parts and distances between holes

Shaving may be completed as a separate operation or it may be incorporated into one station along with punching in a progressive die. A thin-ringed layer of metal is removed by shaving (Fig. 4.12).

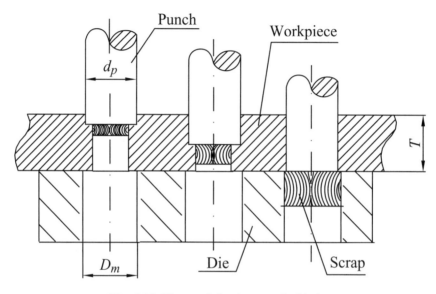

Fig. 4.12 Phases of shaving punched hole.

It is necessary, of course, to provide a small amount of stock of the punched workpiece for subsequent shaving. This amount is:

$$\delta = D - d \tag{4.11}$$

The value of δ is 0.15 to 0.20 mm for a previously punched hole and 0.10 to 0.15 mm for a previously drilled hole.

The diameter of punch in Fig. 4.13, can be calculated from the formula

$$d_{\mathrm{p}} = D + \Delta + i = d + \delta + \Delta + i \tag{4.12}$$

where

D = diameter of hole after shaving

Δ = production tolerance of the hole

i = amount of compensation for tightening of the hole area after shaving

This amount is:

$i = 0.007$ to 0.017 mm for brass

$i = 0.005$ to 0.010 mm for aluminum

$i = 0.008$ to 0.015 mm for low-carbon steel

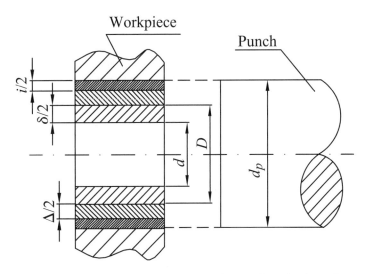

Fig. 4.13 Schematic illustration of an addition for shaving, and production tolerance of the hole.

The diameter of the die needs to be 20 to 30% bigger than the diameter of the punch; that is:

$$D_{\mathrm{m}} = (1.2 \text{ to } 1.3)d_{\mathrm{p}} \tag{4.13}$$

When shaving, the following should be considered:

- Always maintain the recommended close shave clearance for the shave operation.
- One of the natural problems with shaving is keeping the scrap out of the die. For removing this scrap, Arnold recommends the patented Bazooka Bushing, which creates a vacuum force to pull the scrap out of the die button.
- Another problem with shaving in automatic dies is progression control. French notching or trimming both sides of the strip in the die plus adequate piloting should be considered before beginning.

The shave force can be calculated from the formula:

$$F = A \cdot p \tag{4.14}$$

where

A = cross-section of ringed scrap in Fig. 4.13

$A = 0.875[(d + \delta + \Delta + i)^2 - d]$

p = specific pressure

Table 4.5 gives the value for specific pressure $P(MP_\alpha)$.

Table 4.5 Value for specific pressure p(MPa)

Material thickness (mm)	RINGED SCRAP CROSS-SECTION A(mm²)						
	0.05	0.10	0.15	0.20	0.25	0.27	0.30
0.4 to 0.5	1961	1765	1560	1375	1175	1080	980
0.6 to 2.0	2100	1900	1670	1130	1225	1125	1030
2.1 to 3.0	2250	2010	1765	1520	1275	1175	1080

Punching and shaving operations can be done with the same punch at one stroke of the press. For such combinations, different shapes of punches may be used depending on the workpiece material. The gradual punch shown in Fig. 4.14 is used for shaving brass, bronze, and aluminum alloys.

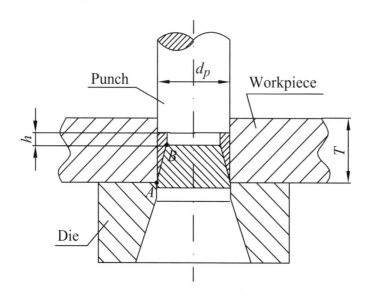

Fig 4.14 Gradual punch for punching and shaving.

The first step of the punch is a smaller dimension, punching the work material, and the second level of punch is a larger dimension, shaving the hole. At the punching stage, the material is shared by plane AB, between the cutting edges of the die and the first punch level.

If the punch is given the shape of a conic section, as shown in Fig 4.15, the process is completed in two phases. In the first, the punch penetrates into the material to a depth of $(0.5$ to $0.75)T$, but the material does not divide. In the second phase, the first crack appears and the scrap begins to slip. This method improves the surface finish of the punched hole. Another advantage of this type of punch is its ability to be sharpened many times.

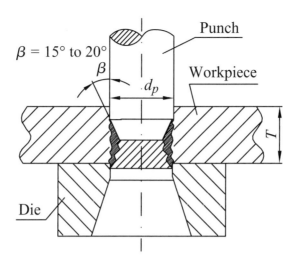

Fig. 4.15 Frustum cone punch for punching and shaving.

In an analogous process of deformation, the material is punched with the conic punch shown in Fig 4.16 but, in this design the punch penetrates the material to a depth of $(0.7$ to $0.9)$ T before the material begins to shear. Two or even three shaves may be required to improve the edge straightness and surface finish. Angle of cone of punch is: $\alpha = 60^0$ to 90^0.

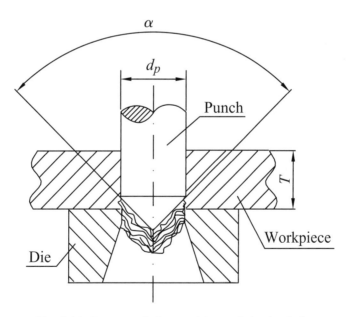

Fig. 4.16 Cone punch for punching and shaving hole.

The progressive punch shown in Fig. 4.17 makes it practical to produce a smooth edge on a work-piece during a single press stroke. The first step of the punch punches a hole, whose diameter is:

$$d_1 = d = D - \delta \ \text{(mm)} \tag{4.15}$$

The second step of the punch uses a sharp edge, and accomplishes the rough shaving of the hole, whose diameter is:

$$d_2 = d + \delta + \Delta + i - 0.01 \text{ (mm)} \tag{4.16}$$

The third step of the punch has rounded edges which perform the final shaving and improvement of the surface finish. The diameter of this step of the punch is:

$$d_3 = d_2 + 0.01 \text{ (mm)} \tag{4.17}$$

A progressive punch is used for shaving holes of diameter $d = (3 \text{ to } 12)$ mm, and material thickness $T < 3$ mm.

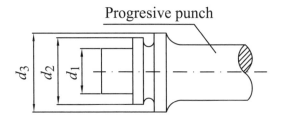

Fig. 4.17 Progressive punch for shaving.

4.5.2 Shaving a Blanked Workpiece

Thin layers of material can be removed from blanked surfaces by a process similar to punching. Figure 4.18 illustrates the shaving process for improving the accuracy of a blanked workpiece. If the workpiece, after shaving, needs to have a diameter D, the punch diameter for blanking operation is:

$$d_p = D + \delta$$

The die diameter for the blanking operation is:

$$d_m = d_p + 2c = D + \delta + 2c$$

where

$\quad\quad D$ = diameter of final workpiece

$\quad\quad c$ = clearance between die and punch

$\quad\quad d$ = amount of material for shaving

The value of an amount d can be found from Table 4.6; it depends on the kind and thickness of material.

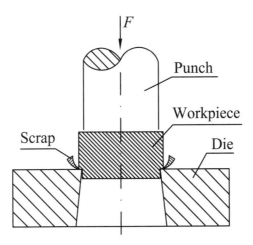

Fig. 4.18 Shaving a blanked workpiece.

Table 4.6 Values for *d* depending on type and thickness of the materials

Material thickness (mm)	Material		
	Copper, aluminum, and low-carbon steel	**Medium-carbon steel**	**High-carbon and alloy steel**
0.5 to 1.4	0.10 to 0.15	0.15 to 0.20	0.15 to 0.25
1.5 to 2.8	0.15 to 0.20	0.20 to 0.25	0.20 to 0.30
3.0 to 3.8	0.20 to 0.25	0.25 to 0.30	0.25 to 0.35
4.0 to 5.2	0.25 to 0.30	0.30 to 0.35	0.30 to 0.40

Example:

A round disk of $d = 76$ mm diameter is to be blanked from a strip of $T = 3$ mm thick, half-hard cold-rolled steel with 40% carbon whose shear strength $\tau_m = 530$ N/mm². Determine:

    a) The appropriate punch and die diameters (d_p and d_m)

    b) blanking force (F)

Solution:

    a) According to equation (4.2), the clearance allowance for half-hard cold-rolled steel is

$$c = \frac{kT\sqrt{\tau_m}}{6.32} = \frac{0.01 \times 3.0\sqrt{530}}{6.32} = 0.109 \text{ mm.}$$

$$\frac{c}{T} \cdot 100 = \frac{0.109}{3} \cdot 100 = 3.65\%$$

According to Table 4.1

$$\frac{2c}{T} \cdot 100 = 9\%; \quad c = \frac{0.09 \cdot 3}{2} = 0.135 \text{ mm};$$

According to Table 4.2

$$2c = 0.21; \quad c = 0.105 \text{ mm}$$

$$\frac{c}{T} \cdot 100 = \frac{0.105}{3} \cdot 100 = 3.5\%$$

The blank is to have a $d = 76$ mm diameter, and die size diameters blank size. Therefore, the die opening diameter is

$$d_m = d = 76 \text{ mm}$$

Punch diameter is:

$$d_p = d_m - 2c = 76 - 2 \times 0.218 = 75.782 \text{ mm}$$

b) To determine the blanking force, we assume that the entire perimeter of the blank is blanked with parallel cutting edge of the punch and die. The length of the cut edge is

$$L = \pi d = 3.14 \times 76 = 238.64 \text{ mm}$$

and the blanking force is

$$F = LT\tau_m = 238.64 \times 3.0 \times 530 = 379,437 \text{ N}$$

The force required of the press F_P is

$$F_p = 1.3F = 1.3 \times 379437 = 493268.8 \text{ N} = 493.269 \text{ kN} \approx 50 \text{ tons.}$$

Five

Bending

5.1 INTRODUCTON

One of the most common processes for sheet-metal forming is bending, which is used not only to form pieces such as L, U, or V- profiles, but also to improve the stiffness of a piece by increasing its moment of inertia. Bending consists of uniformly straining flat sheets or strips of metal around a linear axis, but it also may be used to bend tubes, drawn profiles, bars, and wire. The bending process has the greatest number of applications in the automotive and aircraft industries and for the production of other sheet-metal products. Typical examples of sheet-metal bends are illustrated in Fig. 5.1.

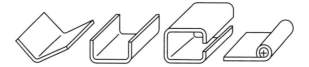

Fig. 5.1 Typical examples of sheet-metal bend parts.

5.2 MECHANICS OF BENDING

The terminology used in the bending process is explained visually in Fig. 5.2. The bend radius R_i is measured on the inner surface of the bend piece. The bend angle u is the angle of the bent piece. The bend allowance is the arc of the neutral bend line.

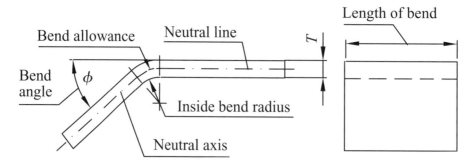

Fig. 5.2 Schematic illustration of terminology used in the bending process.

The length of the bend is the width of the sheet. In bending, the outer fibers of the material are placed in tension and the inner fibers are placed in compression. Theoretically, the strain on the outer and inner fibers is equal in absolute magnitude and is given by the following equation:

$$e_0 = e_t = \frac{1}{\left(2R_i/T\right)+1} \tag{5.1}$$

where

R_i = bend radius

T = material thickness

Experimental research indicates that this formula is more precise for the deformation of the inner fibers of the material e_i than for the deformation of the outer fibers e_0. The deformation in the outer fibers is notably greater, which is why the neutral fibers move to the inner side of the bent piece. The width of the piece on the outer side is smaller and on the inner side is larger than the original width. As R/T decreases, the bend radius becomes smaller, the tensile strain at the outer fibers increases, and the material eventually cracks.

5.3 MOMENT OF BENDING

Suppose we have a long, thin, straight beam having a cross-section *(b × T)* and length *L,* bent into a curve by moments *(M)*. The beam and moments lie in the vertical plane *nxz,* as shown in Fig. 5.3. At a distance *x* from the left end, the deflection of the beam is given by distance *z*. Fig. 5.3b shows, enlarged, two slices *A-B* and *A′-B′* of different lengths *dx,* cut from the beam at location *x*.

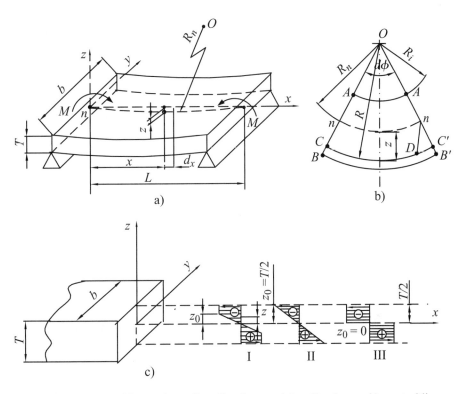

Fig. 5.3 Schematic illustration of bending beam: a) bending beam; b) neutral line; c) bending stress in elastic-plastic zone.

The planes cutting *A-B* and *A-B′* are taken perpendicular to the longitudinal axis *x* of the original straight beam. It is customary to assume that these cross-sections will remain planar and perpendicular to the longitudinal elements of the beam after moments (*M*) are applied (Bernoulli hypothesis). Laboratory experiments have in general verified this assumption. After bending, some of the fibers have been extended (*B-B′*), some have been compressed (*A-A′*), and at one location, called the neutral surface, no change in length has taken place (*n-n*).

The loading of Fig. 5.3 is called pure bending. No shear or tangential stress will exist on the end surfaces *A-B* and *A′-B′*, and the only stress will be σ, acting normally on the surface. An equation can be derived to give the value of this bending stress at any desired distance *z* from the neutral surface.

Let *O* be the center of curvature for slice *n-n* of the deformation beam, *du* the small angle included between the cutting planes, and R_n the radius of curvature. Consider a horizontal element located a distance *z* below the neutral surface. Draw a line *n-D* parallel to *O-B*. The angle *n-O-n* is equal to the angle *D-n-C′* and the following proportional relationship results:

$$\frac{z}{R_n} = \frac{z\,d\varphi}{dx} = \varepsilon \tag{5.2}$$

Because the total deformation of the element $2d\varphi$ divided by the original length dx is the unit deformation or strain, equation (5.2) indicates that the elongation of the element will vary directly with the distance z from a neutral surface.

For a more detailed definition of the stress-to-strain relationship in the bending process, the concept of a reduction in the radius of the neutral curvature (R_r) is useful. This value is the ratio to the bend radius of the neutral surface-to-material thickness:

$$R_r = \frac{R_n}{T} \tag{5.3}$$

where:

    R_r = reduction radius of the neutral curvature surface

Considering the kind and magnitude of stresses that exist during beam bending, as well as the reduction radius (R_r), bending problems can be analyzed in two ways:

    a) Bending in the centrally located inner zone, on both sides of the neutral zone, is a domain of elastic-plastic deformation

    b) Bending in the outlying zones, on both the inside and outside of the bend, is a domain of pure plastic deformation

Bending as a domain of elastic-plastic deformation (Fig. 5.3c) can be considered as a linear stress problem. The true stresses in the bent beam are in the intervals:

$$0 < k < k_m$$

The reduction radius of the neutral surface is in the intervals:

$$5 \leq R_r \leq 200$$

The following events may occur during the bending process:

    a) The core of the beam up to a certain level $(z_0 < T/2)$ may be elastically deformed with both sides of the neutral surface; .but from that level to $z_0 = T/2$ the fibers may be plastically deformed (Fig. 5.3c-I). Assume that:

$$\sigma_e \approx \sigma_{02}$$

    Assume that the material of the beam follows Hooke's law (see Fig. 2.6). Because the strains at the yield point (Y) are very small, the difference between the true and the engineering yield stresses is negligible for metals, and that is:

$$k_0 = YS = \sigma_{02}$$

    This phenomenon is bending in the elastic-plastic domain, because the core of the beam is elastically deformed, and the fibers nearer the outer and inner sides are plastically deformed.

b) The magnitude of the stresses is directly proportional to the fibers' distance from the neutral surface, but the maximum stresses in the inner (*A-A'*) and outer (*B-B'*) fibers are less than the yield stresses. Figure 5.3c-II shows that stresses in the outer and inner fibers are as follows:

$$\sigma < \sigma_{02}$$

This phenomenon is bending in the elastic domain of the material.

c) The stresses may be constant in all cross-sections of the beam and equal to the yield stress (Fig. 5.3c-III). Providing that the material is ideally plastic and does not harden ($k_0 = YS = $ constant), this kind of bending is in the linear-plastic domain. A pure plastic bending of the beam will appear if:

$$R_r < 5$$

In that case, true stresses are in intervals:

$$k_M \leq k < k_f$$

For all cases of bending, Bernoulli's hypothesis concerning the cutting of the planes is in effect.

5.3.1 Moment of Bending in the Elastic-Plastic Domain

The engineering moment of bending in the elastic-plastic domain can be expressed as the sum of the moments of bending in the elastic and plastic zones for the same axis. This sum is given by the general formula,

$$M = YS \left[\frac{2}{z_0} \int_0^z z^2 \, dA + 2 \int_w^{T/2} z \, dA \right]$$

The first segment of this equation is the moment of resistance in the elastic deformation zone with regard to the *y*-axis:

$$W = \frac{2}{z_0} \int_0^{z_0} z^2 \cdot dA$$

The second segment of the equation is the moment of static at the plastic deformation zone with regard to the *y*-axis:

$$S = 2 \int_{z_0}^{T/2} z \cdot dA$$

Therefore, the bending moment in the elastic-plastic domain in the final form is:

$$M = YS(W + S) \tag{5.4}$$

where

  YS = yield strength

  W = moment of resistance

  S = moment of static

For a rectangular cross-section of a beam, the bending moment in the elastic-plastic domain is given by the formula:

$$M = \frac{(YS)b}{12}\left(3T^2 - 4z_0^2\right)$$ (5.4a)

The value of z_0 can be calculated by Hooke's law:

$$YS = E \cdot \varepsilon_0 = E\frac{z_0}{R_n}$$

Therefore,

$$z_0 = \frac{(YS) \cdot R_n}{E}$$

When the expression above is substituted for z_0 (in equation 5.4a), equation 5.4a changes to:

$$M = \frac{(YS) \cdot b}{12}\left[3T^2 - \left(\frac{2(YS) \cdot R_n}{E}\right)^2\right]$$ (5.4b)

Respecting equation (5.3), the bending moment may be expressed as the reduction radius of a curve (R_r):

$$M = (YS)\frac{b \cdot T^2}{12}\left[3 - \left(\frac{2(YS)R_r}{E}\right)^2\right]$$ (5.4c)

However, with bending in the elastic-plastic domain $5 \le R_r \le 200$, the influence of part of the equation $\left(\frac{2 \cdot YS \cdot R_r}{E}\right)^2$ is very slight, and the engineering calculation can be disregarded. Setting aside this part of the equation, we may assume, as a matter of fact, that the entire cross-section of the beam experiences linear-plastic deformation (Fig. 5.3c). Therefore, the moment of the bending beam is loaded by stresses in the linear-plastic domain:

$$M = \left(YS\frac{bT^2}{4}\right)$$ (5.5)

5.3.2 Moment of Bending in the Purely Plastic Domain

The moment of bending in the purely plastic domain for a rectangular cross-section is given by the formula:

$$M = \beta \cdot k\frac{bT^2}{4}$$ (5.6)

where:

β = hardening coefficient of material

k = true strain of material

b = width of beam (length of bending)

T = material thickness

This expression can be simplified to:

$$M = n(UTS)\frac{bT^2}{4}$$

(5.6a)

where

n = correction coefficient hardening of the material (n =1.6 to 1.8)

UTS = ultimate tensile strength of the material

b = width of beam (length of bending)

T = material thickness

5.4 BENDING FORCES

Bending forces can be estimated, if the outer moments of bending and the moments of the inner forces are equal, by assuming that the process is one of a simple bending beam. Thus, the bending force is a function of the strength of the material, the length and thickness of the piece, and the die opening.

5.4.1 Force for U-Die

The bending force for a U-die (Fig. 5.4a) can be generally expressed by the formula:

$$F = \frac{2M}{l}(1 + \sin\varphi)$$

The generally expressed formula is converted for application to sheet metal bending as follows:

$$F = 0.67\frac{(UTS)wT^2}{l}$$

(5.7)

where

UTS = ultimate tensile strength

w = length of bend

l = die opening, $l = R_i + R_k + T$ (see Fig. 5.4)

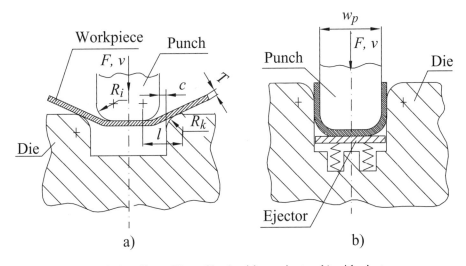

Fig.5.4 Bending a U-profile a) without ejector; b) with ejector.

If the bending is in a die with an ejector (Fig. 5.4b), the bending force needs to increase by about 30 percent so that the total bending force for a U-die is:

$$F_1 = 1.3F \qquad (5.7a)$$

If the bottom of the piece needs to be additionally planished (Fig. 5.5), the force required for bending and planishing is given by the equation:

$$F = p \cdot A \qquad (5.7b)$$

where

p = specific pressure (Table 5.1)

A = area of the bottom

Table 5.1 Value of specific pressure p (MPa)

Material thickness (mm)	Material			
	Aluminum	**Brass**	**Low carbon steel (0.1 to 0.2)%C**	**Steel (0.25 to 0.35)%C**
<3	29.4 to 39.2	58.8 to 78.4	78.4 to 98.0	98.0 to 117.6
3 to 10	49.0 to 58.8	59.8 to 78.4	98.0 to 117.6	117.6 to 147.1

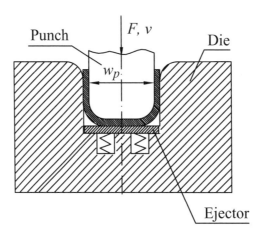

Fig. 5.5 Planished bottom.

Example:

A sheet-metal blank is to be bent, and the bottom of the piece is to be planished in a U-die. The metal has a modulus of elasticity $E = 0.206 \times 10^6$ MPa, $UTS = 579.57$ MPa, the starting blank is $w = 300$ mm wide, material thickness $T = 10$ mm.

Determine the bending force, and force for bottom planishing. A U-die will be used with the following data: punch radius $R_i = 20$ mm, die radius $R_k = 15$ mm, and punch wide $w_p = 200$ mm (Fig. 5.4).

Solution:

Die opening is:

$$l = R_i + R_k + T = 20 + 15 + 10 = 45 \text{ mm} = 0.045 \text{ m}$$

a) Bending force (equation 5.7):

$$F = 0.67 \frac{(UTS)wT^2}{l} = 0.67 \frac{579.57 \cdot 10^6 \times 0.3 \times 0.01^2}{0.045} = 258.9 \text{ kN}$$

b) Bottom planishing force (equation 5.7b):

$$F = pA$$

Specific pressure (Table 5.1):

$$p = 117.6 \text{ MPa} = 117.6 \times 10^6 \text{ Pa}$$

Bottom planishing area:

$$A = (w_p 2R_i)w = (0.2 - 2 \times 0.02) \times 0.3 = 0.048 \text{ m}^2$$
$$F = pA = 117.6 \times 10^6 \times 0.048 = 5644.8 \text{ kN}$$

5.4.2 Forces for a Wiping Die

The bending force for a wiping die (Fig. 5.6) is two times less than for a U-die. It is given by the formula:

$$F = 0.33 \frac{(UTS)wT^2}{l} \tag{5.8}$$

where

$$l = R_t + R_k + T \quad \text{(see Fig. 5.6)}$$

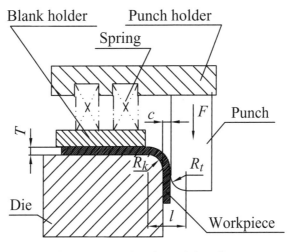

Fig. 5.6 Bending in a wiping die.

5.4.3 Forces for V-Die

Bending V-profiles may be considered as air bending (free bending) (see Fig.5.7) or as coin bending (Fig. 5.8). What exactly do these terms mean? In the beginning phase of bending, the distance between the holds is $\left(l_k - 2R_k\right)$ and the force is applied in the middle, between the holds.

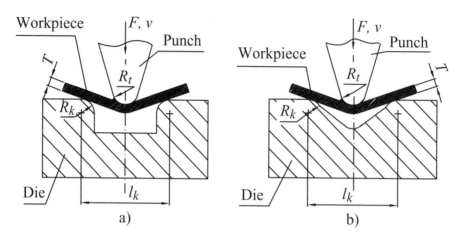

Fig. 5.7 Air bending a) right-angle die profile; b) acute-angle die profile.

The profile of a die for air bending V-profiles can have a right angle, as shown in Fig. 5.7a, or an acute angle, as shown in Fig 5.7b. In this initial phase, the edges of the die with which the workpiece is in contact are rounded at radius R_k. The radius of the punch R_t will always be smaller than the bending radius. The force for air bending a V-profile is given by the formula:

$$F = \frac{4M}{l_k - 2\left(R_k + R_t + T\right)\sin\dfrac{\varphi}{2}}\cos^2\frac{\varphi}{2} \tag{5.9}$$

where

l_k = die opening

φ = bend angle

The coin bend process of the V-profile has four characteristic phases (Fig. 5.8). Phase I is free bending — the distance between the bend points of the die is unchanged and it is equal to l_k. In Phase II, the ends of the workpiece are touching the side surfaces of the die, the bend points of the die are changed, and the bend radius is bigger than the punch radius. In Phase III, the ends of the workpiece are touching the punch. Between Phase III and Phase IV, the workpiece is actually being bent in the opposite direction from Phase I and II (negative springback). When the workpiece is touching the die and the punch on all surfaces, the bend radius and the punch radius are equal; phase IV is then terminated, and bending the workpiece is completed.

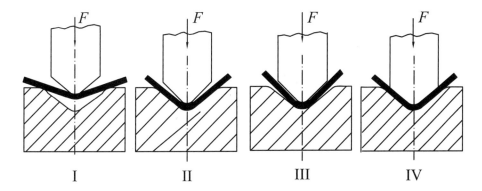

Fig. 5.8 V-profile bending phases.

It is usually necessary to flatten the bottom bend area between the tip of the punch and the die surface in order to avoid springback. At the moment of completing phase IV, it is advisable to increase the force for a final reinforcement of the bend and completion of the bottoming operation. The force necessary for this final reinforcement is given by the formula:

$$F_1 = 2 \cdot p \cdot b \cdot c \cdot \cos\frac{\varphi}{2} \tag{5.9a}$$

where

p = specific pressure (Table 5.1)

b = contact length (width of the workpiece)

c = length of the straight end of the workpiece

The relationship between the bend forces and the punch travels is shown in Fig. 5.9. Air bending (interval-OG) has three parts. The first part is elastic deformation (OE). In the second, the force is mostly constant (EF), and in the third, the force decreases because of material slip (FG). After that, the force again increases to a definitive point. The workpiece is bent (GH). If the workpiece needs to be bottomed, the force very quickly increases (HM).

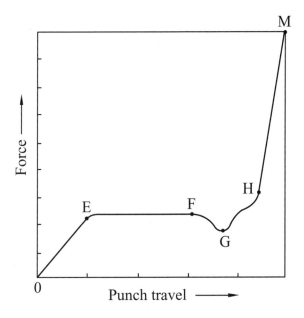

Fig. 5.9 Typical load-punch travel curve for coin bending process.

Example:

A sheet-metal blank is to be air bent and a final reinforcement of the bend is to be made in a V-die. The material has $UTS = 0.6864$ MPa (aluminum), the starting blank is $w = 400$ mm wide, the material thickness is $T = 2$ mm, the length of the straight end of the piece is $c = 50$ mm, and the bend angle is $\varphi = 120^0$.

Determine the air bending force, and the force necessary for the final workpiece bend reinforcement. The same V-die will be used for both the air bending and reinforcement operations, with the die possessing the following parameters: punch radius $R_t = R_i = 4$ mm, die radius $R_k = 3$ mm, die opening $l_k = 89$ mm (Fig. 5.7b).

Solution:

a) Air bending force (Equation 5.9):

$$F = \frac{4M}{l_k - 2(R_k + R_i + T)\sin\dfrac{\varphi}{2}}\cos^2\frac{\varphi}{2}$$

Moment of bending (Equation 5.6a):

$$M = n(UTS)\frac{wT^2}{4} = 1.8 \times 686465.5\frac{0.4 \times 0.002^2}{4} = 0.4942 \text{ Nm}$$

$$F = \frac{4 \times 0.4942}{0.089 - 2(0.003 + 0.004 + 0.002)0.866}0.25 = 6.732 \text{ N}$$

Force necessary for final reinforcement (Equation 5.9a):

$$F_1 = 2p \cdot w \cdot c \cdot \cos\frac{\varphi}{2} = 2 \times 29.4 \times 10^6 \times 0.4 \times 0.005 \cdot 0.5 = 58.8 \text{ kN}$$

5.4.4 Curling

Two examples of curling are shown in Fig. 5.10 and Fig. 5.11. Curling gives stiffness to the workpiece by increasing the moment of inertia at the ends, and providing smooth rounded edges. In the first example in Fig. 5.10, the edge of the sheet metal is bent into the cavity of a punch.

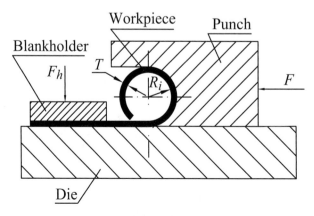

Fig. 5.10 Curling process.

In the second example in Fig. 5.11, the circular edge of the initial deep-drawn workpiece is curled by a tool that incorporates a cavity punch.

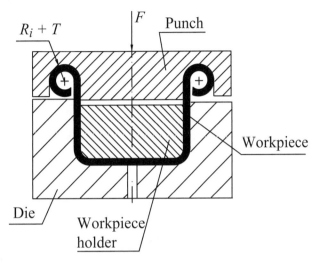

Fig. 5.11 Circular edge curling.

The curling force is given by the equation:

$$F = \frac{M}{R_i + 0.5T}$$ (5.10)

where

M = moment of bending

R_i = inside curling radius

T = material thickness

Example:

Define the curling force for the workpiece shown in Fig. 5.11. Assume:

Diameter	$D = 400$ mm
Material thickness	$T = 1.2$ mm
Inner radius	$R_i = 1.2$ mm
The ultimate tensile strength	$UTS = 176.5$ MPa

Solution:

$$F = \frac{M}{R_i + 0.5T}$$

$$M = n(UTS)\frac{bT^2}{4} = 1.8(176.5 \times 10^6)\frac{3.14 \times 0.4 \times 0.0012^2}{4} = 143.651 \text{ Nm}$$

$$F = \frac{143.651}{0.003 + 0.5 \times 0.0012} = 39.903 \text{ kN}$$

Known bend and curl forces often are not so important for the process because, very often, the maximum force of the press machine is greater than the bending or curling force. However, knowing the magnitude of these forces is necessary for a definition of the blankholder forces. Because of the phenomenon of material fatigue of the blank springs, these forces need to be 30 to 50 percent greater than the bending or the curling forces.

5.4.5 Three-Roll Forming

For bending differently shaped cylinders (plain round, corrugated round, flattened, elliptical, etc.) or truncated cones of sheet metal, the three-roll forming process is used. Depending upon such variables as the composition of the work metal, machine capability, or part size, the shape may be formed in a single pass or a series of passes. Fig. 5.12 illustrates the basic setup for three-roll forming on pyramid-type machines.

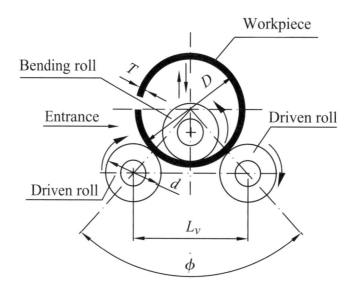

Fig. 5.12 Three-roll bending.

The two lower rolls on pyramid-type machines are driven, and the adjustable top roll serves as an idler and is rotated by friction with the workpiece.

In most set-ups, short curved sections of circular works are pre-formed on the ends of the metal workpiece in a press brake or on hydraulic press. Otherwise, the workpiece would be ends that, instead of being curved, would be straight. In the process described above, the radius of the bend allowance is much greater than the material thickness of the workpiece; under these conditions, the bending is entirely in the elastic-plastic domain.

To achieve permanent deformation in the other and inner fibers of the material, the following relationship must apply:

$$\frac{D}{T} < \frac{E}{YS} + 1 \tag{5.11}$$

Otherwise, the workpiece, instead of being curved, will be straight after unloading. The bending force on the upper roll is given by the formula:

$$F = YS \frac{b}{D-T} \left[T^2 - \frac{YS(D-T)^2}{3E^2} \right] \operatorname{ctg} \frac{\varphi}{2} \tag{5.12}$$

where

      D = outer diameter of the workpiece

      b = length of bend

      T = material thickness

      YS = yield stress

      E = modulus of elasticity,

      ϕ = bend angle

The bend angle can be calculated from the geometric ratio in Fig. 5.12 and is given by the formula:

$$\varphi = 2\arcsin\frac{L_v}{D+d}$$

where

      L_v = distance between lower rolls

      d = lower rolls diameter

Example:

Define the bending force for the cylinder for which outer diameter is D = 2000 mm, wide b = 1000 mm, and material thickness T = 10 mm.

    The metal has a modulus of elasticity E = 210,843 MPa; UTS = 441.3 MPa; and YS = 235.36 MPa.

    The cylinder is bent on a machine with three-rollers. The diameter of rolls is d = 600 mm, and the distance between the lower rolls is Lv = 200 mm (Fig.5.12).

Solution:

To achieve permanent deformation in the outer and inner fibers of the material, the relationship must be apply (equation 5.11).

$$\frac{D}{T} < \frac{E}{YS} + 1 = \frac{2}{0.010} < \frac{210843}{235.36} + 1 = 200 < 897$$

Hence the relationship is satisfied.

    Bending force on the upper roll (equation 5.12)

$$F = YS\frac{b}{D-T}\left[T^2 - \frac{YS(D-T)^2}{3E^2}\right]\text{ctg}\frac{\varphi}{2}$$

Bend angle:

$$\varphi = 2\arcsin\frac{L_v}{D+d} = \frac{200}{2000+200} = 31° \; 36' \;...; \; \frac{\varphi}{2} = 15° \; 48'$$

$$F = YS\frac{b}{D-T}\left[T^2 - \frac{YS(D-T)^2}{3E^2}\right]\text{ctg}\frac{\varphi}{2}$$

$$F = 235.36\frac{1}{2-0.01}\left[0.01^2 - \frac{235.36^2(2-0.01)^2}{3\times210843^2}\right]\times3.534$$

$$F = 118.3\times98.36\times3.534 = 41.1563\text{ kN}$$

5.5 BEND RADIUS

One of the most important factors that influence the quality of a bent workpiece is the bend radius R_i (see Fig. 5.3), which must be within defined limits. The bend radius is the inside radius of a bent workpiece.

5.5.1 Minimum Bend Radius

The minimum bend radius is usually determined by how much outer surface fracture is acceptable. However, many other factors may limit the bend radius. For instance, wrinkling of the inner bend surface may be of concern if it occurs before initiation of fracture on the outer surface. In developing a description of the minimum bend radius, it is necessary to have some knowledge of the amount of *strain* imposed and the material *ductility*. We have a good definition of strain, but the term ductility is vague. It is necessary to have a quantitative measurement of the amount of deformation that the material can undergo before fracture. As with most mechanical properties, fracture strain can be obtained from tensile testing. There may be no need to run a bending test if tensile test data are available, which they usually are.

The strain of certain fibers at distance z from the neutral surface is defined by formula (5.2).

$$\varepsilon = \frac{z}{R_n} = \frac{R-R_n}{R_n}$$

The greatest tensile strain appears in the outer fibers: $R = R_i + T$ (see Fig. 5.3). When $R_n = R_i + T/2$, tensile strain can be calculated by the following:

$$e = \frac{1}{(2R_i/T)+1} \tag{5.13}$$

If the strain at which the cracks in the outer fibers appear is defined as e_f, and the minimum bend radius, which causes these strains, as R_{min}, then:

$$R_{min} = \frac{T}{2}\left[\frac{1}{e_f}-1\right] \tag{5.14}$$

It is apparent from equation (5.13) that as the R_i/T ratio decreases, the bend radius becomes smaller, the tensile strain on the outer fibers increases, and the material may crack after a certain strain is reached. The minimum radius to which a workpiece can be bent safely is normally expressed in terms of the material thickness and is given by the following formula:

$$R_{i(min)} = c \cdot T \qquad (5.14a)$$

The coefficient c for a variety of materials has been determined experimentally, and some typical results are given in Table 5.2.

Table 5.2 Values of the coefficient c

MATERIAL	CONDITION	
	Soft	**Hard**
Low carbon steel	0.5	3.0
Low alloy steel	0.5	4.0
Austenitic stainless steel	0.5	4.0
Aluminum	0.0	1.2
Aluminum alloy series 2000	1.5	6.0
" series 3000	0.8	3.0
" series 4000	0.8	3.0
" series 5000	1.0	5.0
Copper	0.25	4.0
Bronze	0.4	2.0
Titanium	0.7	3.0
Titanium alloy	2.5	4.0

The bendability of a material may be increased by techniques such a applying compressive forces in the plane of the sheet during bending to minimize tensile stress in the outer fibers of the bend area, or increasing tensile reduction of the area by heating. If the length of the bend increases, the state of stress at the outer fibers changes from uniaxial stress to biaxial stress, which reduces the ductility of the material. Therefore, as the length increases, the minimum bend radius increases. Bendability decreases with rough edges because rough edges form points of stress concentration. Anisotropy of the sheet metal is also an important factor in bendability. If the bending operation takes place parallel to the direction of rolling, separations will occur and cracking will develop, as shown in Fig. 5.13a.

If bending takes place at right angles to the rolling direction of the sheet metal, there should be no cracks, as shown in Fig. 5.13b. In bending such a sheet or strip, caution should be used in cutting the blank from the rolled sheet in the proper direction, although that may not always be possible in practice.

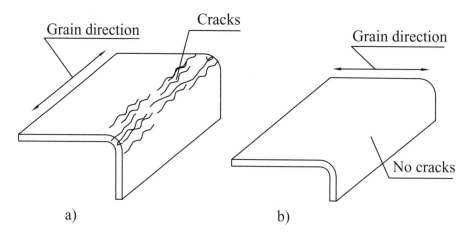

Fig. 5.13 Relationship between grain direction and cracking on bent parts: a) cracking results when the direction of bending is parallel to the original rolling direction of the sheet; b) bending at an angle to the original rolling direction of the sheet will tend to avoid cracking.

5.5.2 Maximum Bend Radius

If it is considered that the bend radius of the neutral surface is $R_n = R_i + \dfrac{T}{2}$, the engineering strain rate is:

$$e = \frac{(R_i + T) - (R_i + T/2)}{R_i + T/2} = \frac{T/2}{R_i + T/2}$$

Using a large radius for bending $(R_i > T)$ means that the expression $T/2$ in the divider will be of very small magnitude with regard to R_i, and may be ignores so:

$$e = \frac{T}{2R_i}, \quad \text{or} \quad R_i = \frac{T}{2e} = \frac{T}{2} \cdot \frac{E}{\sigma}$$

To achieve permanent plastic deformation in the outer fibers of the bent workpiece, the maximum bend radius must be:

$$R_{i\,(max)} \le \frac{TE}{2YS} \tag{5.15}$$

Therefore, the bend radius needs to be:

$$R_{i\,(min)} \le R_i \le R_{i\,(max)} \tag{5.15a}$$

If this relationship is not satisfied, then one of two results may ensue:

a) for $R_i < R_{i\,(max)} = c \cdot T$, cracks will develop on an outer side of the bent workpiece.

b) for $R_i > R_{(max)} = TE/2(YS)$, permanent plastic deformation will not be achieved in the bent work piece, and after unloading, the workpiece will experience elastic recovery (springback).

Example:

Check the maximum bend radius of the U-profile in Fig. 5.14, using steel as the material with:

$UTS = 750$ MPa

$YS = 620$ MPa

$T = 1$ mm

$E = 215000$ MPa

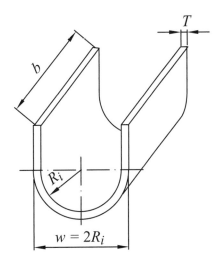

Fig. 5.14 U-profile bend.

Solution:

$$R_{i\,(max)} = \frac{TE}{2(YS)} = \frac{1 \times 215000}{2 \times 620} = 173.38 \text{ mm}$$

$$w = 2R_{i\,(max)} = 2 \times 173.38 = 346.76 \text{ mm}$$

The U-profile cannot be bent because the radius in Fig. 5.14, $R_i = 400$ mm is bigger than the maximum bend radius, $R_{i(max)} = 173.38$ mm, and

$$w_{max} = 2R_{i\,(max)} = 346.76 \text{ mm} < 400 \text{ mm}$$

5.6 BEND ALLOWANCE

The bend allowance is the length of the arc of the neutral bending line for a given degree value. For a large inner bend radius (R_i), the neutral bending line position stays approximately at the mid-thickness of the material. A large bend radius is generally considered to be larger than 5 times the material thickness ($R_i > 5T$). For a smaller bend radius, the neutral bending line shifts toward the inside bend surface. Also, because work material volume is constant in plastic deformation, the blank gets thinner in this elongated zone. There is also a contraction of the workpiece width, but this is usually negligible when the workpiece width is at least 10 times the thickness

($b > 10T$). The amount of neutral bend line shift depends on the inner bend radius. Table 5.3 gives the coefficient (ξ), which determines the amount of the neutral bend line shift depending on the ratio of the inner bend radius (R_i) to the material thickness (T). The general equation for the length of arc at the neutral bending line is given by:

$$L_n = \frac{\pi\varphi}{180} R_n \qquad (5.16)$$

where

φ = bend angle

R_n = bend allowance radius

Table 5.3 Values of coefficient ξ*

Ri/T	0.1	0.2	0.3	0.4	0.5	0.8	10.	1.5	2.0	3.0	4.0	5.0	10.0
ξ	0.23	0.29	0.32	0.35	0.37	0.40	0.41	0.44	0.45	0.46	0.47	0.48	0.50

*source: Sprovochnik Metallist Vol.4

The bend allowance radius (R_n) is the arc radius of the neutral bending line between the form tangency, as shown in Fig. 5.15.

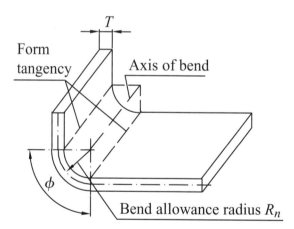

Fig. 5.15 Schematic illustration of terminology: form tangency; Axis of bend, and bend allowance radius.

The bend allowance radius (R_n) for various values of a variety of inner bend radii has been determined theoretically and experimentally. According to the theory by R. Hill, the strains at the borderline of the zone of compression and tension must be zero; then:

$$\frac{R_n}{R_i} = \frac{R_0}{R_n}$$

Therefore, the bend allowance radius is:

$$R_n = \sqrt{R_0 \cdot R_i} \qquad (5.17)$$

where

   R_0 = outer bend radius

   R_i = inner bend radius

The second method for defining the bend allowance radius is given by the following formula:

$$R_n = \xi \cdot T + R_i \tag{5.18}$$

The final equations for calculating bend allowance are:

$$L_n = \frac{\pi\varphi}{180}\left(\xi T + R_i\right)0.017453\varphi\left(\xi T + R_i\right) \text{ or} \tag{5.19}$$

$$L_n = \frac{\pi\varphi}{180}\sqrt{R_0 \cdot R_i} = 0.017453\varphi\left(\sqrt{R_0 \cdot R_i}\right)$$

The majority of bend angles are of 90°, as shown in Fig. 5.15, so it is useful to state the equation for bends. This equation is:

$$L_n = \frac{90\pi}{180}\left(\xi T + R_i\right) = 1.5708\left(\xi T + R_i\right) \tag{5.19a}$$

The length of the blank prior to bending is shown in Fig 5.16:

$$L = L_1 + L_n + L_2 \tag{5.20}$$

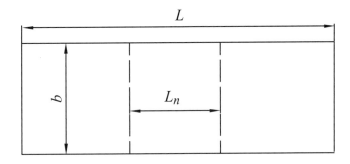

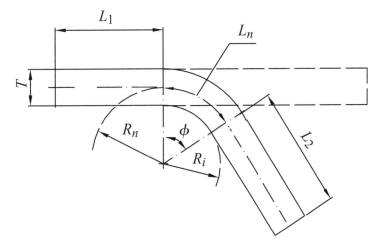

Fig. 5.16 Schematic illustration of pre-bend length.

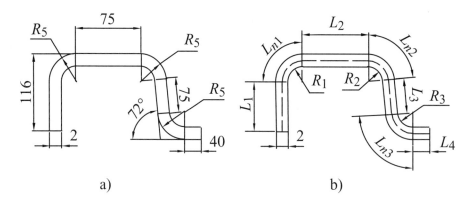

Fig. 5.17 Example developed length of the part.

Example:

Calculate the pre-bend length of the blank shown in Fig. 5.17a.

Solution:

The first thing that must be done is to redimension the drawing so that all dimensions are shown to be internal. This is done in Fig. 5.17b.

The internal radii R_1; R_2; and R_3 are:

$$R_1 = R_2 = R_3 = 5 - 2 = 3 \text{ mm}$$

The length of L_1 is:

$$L_1 = 116 - 5 = 111 \text{ mm}$$

The lengths of the legs L_2, L_3, and L_4 are unchanged, and they are:

$$L_2 = 75 \text{ mm}, L_3 = 75 \text{ mm, and } L_4 = 40 \text{ mm}$$

Because $R_1 = R_2 = R_3 = R_i$, then:

$$\frac{R_i}{T} = \frac{3}{2} = 1.5$$

From Table 5.3 for value $R_i/T = 1.5$ value $\xi = 0.44$, the length of arc L_{n1} is:

$$L_{n1} = \frac{90\pi}{180}\left(\xi T + R_i\right) = 1.5708\left(0.44 \times 2 + 3\right) = 6.09 \text{ mm}$$

The length of L_{n2}

$$L_{n2} = \frac{\pi\varphi}{180}\left(\xi T + R_i\right) = 0.017453\ (72)\left(0.44 \times 2 + 3\right) = 4.94 \text{ mm}$$

$$L_{n3} = \frac{\pi\varphi}{180}\left(\xi T + R_i\right) = 0.017453\ (72)\left(0.44 \times 2 + 3\right) = 4.94 \text{ mm}$$

The developed length is the sum of all these lengths in the neutral bending line. It is:

$$L = L_1 + L_{n1} + L_2 + L_{n2} + L_3 + L_{n3} + L_4$$

$$L = 111 + 6.09 + 75 + 4.94 + 75 + 4.94 + 40 = 117.6 \text{ mm}$$

5.7 SPRINGBACK

Every plastic deformation is followed by elastic recovery. As a consequence of this phenomenon, changes occur in the dimensions of the plastic-deformed workpiece upon removing the load.

Permanent deformation (ε_t) is expressed as the difference between the plastic (ε_{pl}) and the elastic deformation (ε_e) formations:

$$\varepsilon_t = \varepsilon_{pl} - \varepsilon_e$$

When a workpiece is loaded, it will have the following characteristic dimensions as a consequence of plastic deformation (see Fig. 5.18):

- bend radius (R_i)
- bend angle ($\varphi_i = 180^0 - \alpha_1$)
- profile angle (α_1)

All workpiece materials have a finite modulus of elasticity, so each will undergo a certain elastic recovery upon unloading. In bending, this recovery is known as springback (Fig. 5.18). The final dimensions of the workpiece after being unloaded are:

- Bend radius (R_f)
- Profile angle (a_2)
- Bend angle ($\varphi_f = 180^\circ - a_2$)

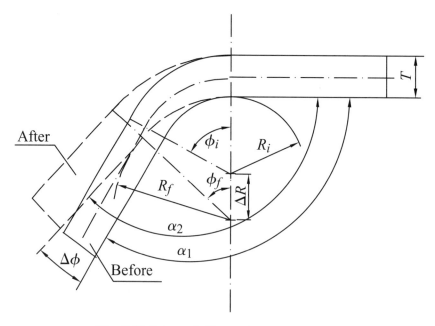

Fig. 5.18 Schematic illustration of springback.

The final angle after springback is smaller $\left(\varphi_f < \varphi_i\right)$ and the final bend radius is larger $\left(R_f > R_i\right)$ than before.

There are two ways to understand and compensate for springback. One is to obtain or develop a predictive model of the amount of springback (which has been proven experimentally). The other way is to define a quantity to describe the amount of springback. A quantity characterizing springback is the springback factor (K_s), which is determined as follows:

The bend allowance of the neutral line (L_n) is the same before and after bending, so the following relationship is obtained by the formula:

$$L_n = \left(R_i + \frac{T}{2} \right) \varphi_i = \left(R_f + \frac{T}{2} \right) \varphi_f$$

From this relationship, the springback factor is:

$$K_s = \frac{R_i + \dfrac{T}{2}}{R_f + \dfrac{T}{2}} = \frac{\dfrac{2R_i}{T} + 1}{\dfrac{2R_f}{T} + 1} = \frac{\varphi_f}{\varphi_i} = \frac{180^0 - \alpha_2}{180^0 - \alpha_1} \tag{5.21}$$

The springback factor (K_s) depends only on the R/T ratio. A springback factor of $K_s = 1$ indicates no springback, and $K_s = 0$ indicates complete elastic recovery. To estimate springback, an approximate formula has been developed in terms of the radii R_i and R_f as follows:

$$\frac{R_i}{R_f} = 4 \left(\frac{R_i(YS)}{ET} \right)^3 - 3 \left(\frac{R_i(YS)}{ET} \right) + 1 \tag{5.22}$$

Values of the springback factor for similar materials are given in Table 5.4.

Table 5.4 Values of springback factor K_s

R_f/T	1.0	1.6	2.5	4.0	6.3	10.0	25.0
Material (AISI)	Springback factor (K_s)						
2024-T	0.92	0.905	0.88	0.85	0.80	0.70	0.35
7075-0 & 2024-0	0.98	0.98	0.98	0.98	0.975	0.97	0.945
7075-T	0.935	0.93	0.925	0.915	0.88	0.85	0.748
1100-0	0.99	0.99	0.99	0.99	0.98	0.97	0.943

In V-die bending, it is possible for the material to exhibit negative springback. This condition is caused by the nature of deformation as the punch completes the bending operation. Negative springback does not occur in air bending (free bending) because of the lack of constraints in a V-die.

In practice, compensation for springback is accomplished by:

a) Overbending the workpiece. In practice, a 2% to 8% addition to the angle of bend is sufficient allowance for spring back. In bending steel parts, the addition to the angle is less than in bending aluminum-alloy parts.

Flanges with a curved surface require greater stress to bend the workpiece and do not spring back as much as straight-line bends.

b) Bottoming (coining the bend area). Bottoming is intensive impact-type loading applied at the bottom of the workpiece by subjecting it to high localized stresses between the tip of the punch and the die surface. A problem with this method of handling springback is that the deformation is not well controlled. Variations in the bend, the blank thickness, and die and die and punch geometries produce different degrees of deformation.

c) Stretch bending. The workpiece is subjected to tension while being bent in a die.

Example:

Assume the bend V-profile will comply with the following data:

- profile angle $\alpha_2 = 108^0$
- bend radius $R_f = 8$ mm
- material 7075-T
- material thickness $T = 2$ mm

Calculate the nose punch radius (R_t) and the punch angle (α_t).

Solution:

The value of: $\dfrac{R_f}{T} = \dfrac{8}{2} = 4$

The value of K_s found from Table 5.4, is: $K_s = 0.915$. The value of R_i from equation (5.21) is:

$$R_i = K_s\left(R_f + \frac{T}{2}\right) - \frac{T}{2} = 0.915(8-1) - 1 = 5.405 \text{ mm}$$

The bend angle is:

$$\varphi_f = 180^0 \alpha_2 = 180^0 - 108^0 = 72^0$$

The punch angle is:

$$\alpha_i = \alpha_1 = 180^0 - \varphi_i = 180^0 - \frac{\varphi_f}{K_s} = 180^0 - \frac{72^0}{0.915} = 101.13^0$$

Values of springback are:

$$\Delta R = R_f - R_i = 8 - 5.405 = 2.595 \text{ mm}$$
$$\Delta \alpha = \alpha_2 - \alpha_1 = 108 = 101.13 = 6.87^0$$

5.8 CLEARANCE

For a definition of the clearance c between the punch and die during the bending process, as shown in Fig. 5.19, it is necessary to know the tolerance of the material at a given thickness, which is given by this formula:

$$c = T_{max} - 0.1 \tag{5.23}$$

The effect of adequate clearance is a smooth pulling of the material into the die. When clearance is too small, the material tends to be sheared rather than bent.

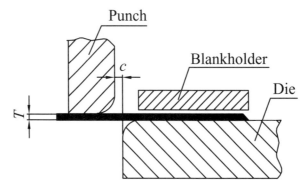

Fig. 5.19 Clearance c between punch and die.

Six

Deep Drawing

6.1 INTRODUCTION

The drawing of metal or "deep drawing" is the process by which a punch is used to force sheet metal to flow between the surfaces of a punch a die. A flat sheet is formed into a cylindrical–, conic–, or box–shaped part. With this process, it is possible to produce a final workpiece using minimal operations and generating minimal scrap that can be assembled without further operations.

The development of specific methods for the deep drawing process has paralleled general technological development, especially in the automotive and aircraft industries. However, this process has a broad application for the production of parts of different shapes and different dimensions for other products, ranging from very small pieces in the electrical and electronic industries to dimensions of several meters in other branches of industry.

Deep drawing is popular because of its rapid press cycle times. Complex axisymmetric geometries, and certain non–axisymmetric geometries, can be produced with a few operations, using relatively non–technical labor.

From the functional standpoint, the deep drawing metal–forming process produces high–strength and light-weight parts as well as geometries unattainable with some other manufacturing processes. There are two deep drawing processes:

- deep drawing without a reduction in the thickness of the workpiece material (pure drawing)
- deep drawing with a reduction in the thickness of the workpiece material (ironing)

A schematic illustration of these deep drawing processes is shown in Fig. 6.1. From the illustration of the deep drawing process without a reduction in the thickness of the workpiece material (Fig.6.1a), it is clear that the basic tools for deep drawing are the punch, the drawing die ring, and the blank holder. In the ironing process (Fig. 6.1b), a previous deep drawn–cup is drawn through one or more ironing rings with a mowing punch to reduce the wall thickness of cup.

Deep drawing is one of the most widely used sheet metal working processes and is used to produce cup–shaped components at a very high rate. Cup drawing, besides its importance as a forming process, also serves as a basic test for sheet metal formability. Typical products are pots and pans, containers of all shapes, sinks, beverage cans, and automobile and aircraft panels.

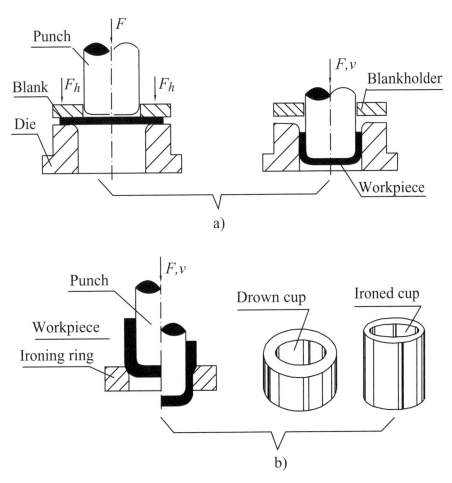

Fig. 6.1 Schematic illustration of deep drawing process: a) drawing without reducing thickness (start of operation and near end of stroke); b) drawing with reducing thickness–ironing (cup before ironing and cup after ironing).

6.2 MECHANICS OF DEEP DRAWING

Deep drawing is the metal forming process by which a flat sheet of metal is cold drawn or formed by a mechanical or hydraulic press into a seamless shell. As the material is drawn into the die by the punch, it flows into a three–dimensional shape. The blank is held in place with a blank holder using a certain force. High compressive stresses act upon the metal, which, without the offsetting effect of a blank holder, would result in a severely wrinkled workpiece.

Wrinkling is one of the major defects in deep drawing; it can damage the dies and adversely affect part assembly and function. The prediction and prevention of wrinkling are very important. There are a number of different analytical and experimental approaches that can help to predict and prevent flange wrinkling. One of them is the finite element method (FEM). However, an explanation of this method falls outside the scope of this book, so it is not explained here.

There are many important variables in the deep drawing process. They can be classified as:

- material and friction factors
- tooling and equipment factors

Important material properties such as the strain hardening coefficient (n) and normal anisotropy (R) affect the deep drawing operation. Friction and lubrication at the punch, die, and workpiece interfaces are very important to obtain a successful deep drawing process. A schematic illustration of the significant variables in the deep drawing process is shown in Fig. 6.2.

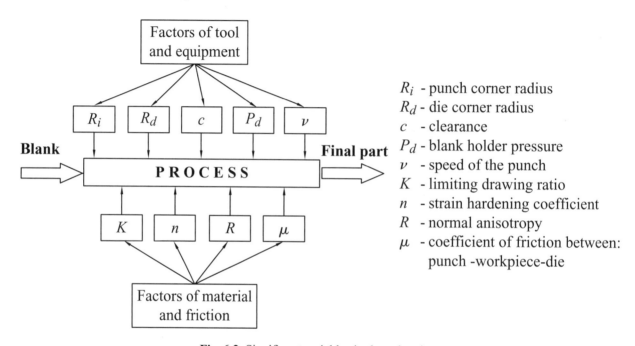

R_i - punch corner radius
R_d - die corner radius
c - clearance
P_d - blank holder pressure
v - speed of the punch
K - limiting drawing ratio
n - strain hardening coefficient
R - normal anisotropy
μ - coefficient of friction between:
punch -workpiece-die

Fig. 6.2 Significant variables in deep drawing.

Unlike bending operations, in which metal is plastically deformed in a relatively small area, drawing operations impose plastic deformation over large areas. Not only are large areas of the forming workpiece being deformed, but the stress states are different in different regions of the part. As a starting point, consider what appear to be three zones undergoing different types of deformation:

- the flat portion of the blank that has not yet entered the die cavity (the flange)
- the portion of the blank being drawn into the die cavity (the wall)
- the zone of contact between the punch and the blank (bottom)

The radial tensile stress is due to the blank being pulled into the female die; the compressive stress, normal in the blank sheet, is due to the blank holder pressure. The punch transmits force F to the bottom of the cup, so the part of the blank that is formed into the bottom of the vessel is subjected to radial and tangential tensile stresses. From the bottom, the punch transmits the force through the walls of the cup to the flange. In this stressed state, the walls tend to elongate in the longitudinal direction. Elongation causes the cup wall to thin, which—if it is excessive—can cause the workpiece to tear. Fig. 6.3a illustrates the fracture of a cup in deep drawing caused by too small a die radius R_d, and Fig. 6.3b shows the fracture caused by too small a punch corner radius R_i. Fracture can also result from high longitudinal tensile stresses in the cup due to a high ratio of blank diameter to punch diameter.

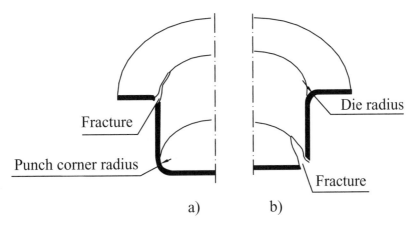

Fig. 6.3 Fracture of a cup in deep drawing: a) caused by too small a die radius; b) caused by too small a punch radius.

The tensile hoop stress on the wall indicates that the cup may be tight on the punch because of its contraction due to the tensile stresses in the cup wall. If drawing is done without blank holder pressure, the radial tensile stresses can lead to compressive hoop stress on the flange. It is these hoop stresses that tend to cause the flange to wrinkle during drawing. Also note that, in pure drawing, the flange tends to increase in thickness as it moves toward the die cavity because its diameter is being reduced. Parts made by deep drawing usually require several successive draws. One or more annealing operations may be required to reduce work hardening by restoring the ductile grain structure. The number of successive draws required is a function of the ratio of the part height h to the part diameter d, and is given by this formula:

$$n = \frac{h}{d} \tag{6.1}$$

where

n = number of draws

h = part height

d = part diameter

The value of n for the cylindrical cup draw is given in Table 6.1.

Table 6.1 Numbers of draws (n) for a cylindrical cup draw

h/d	< 0.6	0.6 to 1.4	1.4 to 2.5	2.5 to 4.0	4.0 to 7.0	7.0 to 12.0
n	1	2	3	4	5	6

6.2.1 Deep Drawability

In deep drawing, deformation may be expressed in four ways, thus:

$$\varepsilon = \frac{D - d_s}{D}; \quad m = \frac{d_s}{D}; \quad K = \frac{D}{d_s}; \quad \varphi = \ln \frac{D}{d_s} \tag{6.2}$$

The relationship between these equations is:

$$\varepsilon = \frac{D-d_s}{D} = 1 - m = \frac{K-1}{K} = \frac{e^{\varphi-1}}{e^{\varphi}} \qquad (6.2a)$$

The ratio of the mean diameter *ds* of the drawn cup to the blank diameter *D* is known as the drawing ratio *m*, and is given by:

$$m = \frac{d_s}{D} = 1 - \varepsilon = \frac{1}{K} = \frac{1}{e^{\varphi}} \qquad (6.2b)$$

Very often, deep deformability is expressed as the reciprocal of the drawing ratio *m*. This value *K* is known as the limit of the drawing ratio:

$$K = \frac{D}{d_s} = \frac{1}{m} = e^{\varphi} \qquad (6.2c)$$

where

D = the blank diameter

d_s = the mean diameter of the drawn cup

m = the drawing ratio

K = the limit of the drawing ratio

The values of the drawing ratio for the first and succeeding operations is given by:

$$m_1 = \frac{d_{s1}}{D}; \quad m_2 = \frac{d_{s2}}{d_{s1}}; \quad m_3 = \frac{d_{s3}}{d_{s2}}; \dots m_n = \frac{d_{sn}}{d_{sn-1}} \qquad (6.3)$$

The magnitude of these ratios determines the following parameters:

- the stresses and forces of the deep drawing process
- the number of successive draws
- the blank holder force
- the quality of the final drawn parts

In view of the complex interaction of factors, certain guidelines have been established for a minimum value of the drawing ratio. The relative thickness of material is the most important and may be calculated from:

$$T_r = \frac{T}{D} 100\% \qquad (6.4)$$

As the relative thickness of the material T_r becomes greater, the drawing ratio *m* becomes more favorable. Table 6.2 provides an optimal drawing ratio for a cylindrical cup without a flange.

Table 6.2 Optimal ratio *m* for drawing a cylindrical cup without flange

Ratio of drawing *m*	Relative thickness of the material $T_r = \dfrac{T}{D} 100\%$					
	2.0–1.5	**1.5–1.0**	**1.0–0.6**	**0.6–0.3**	**0.3–0.15**	**0.15 – 0.08**
m_1	0.48–0.50	0.50–0.53	0.53–0.55	0.55–0.58	0.58–0.60	0.60–0.63
m_2	0.73–0.75	0.75–0.76	0.76–0.78	0.78–0.79	0.79–0.80	0.80–0.82
m_3	0.76–0.78	0.78–0.79	0.79–0.80	0.81–0.82	0.81–0.82	0.80–0.84
m_4	0.78–0.80	0.80–0.81	0.81–0.82	0.80–0.83	0.83–0.85	0.85–0.86
m_5	0.80–0.82	0.82–0.84	0.84–0.85	0.85–0.86	0.86–0.87	0.87–0.88

6.3 FORCES

The first deep drawing operation is not a steady–state process. The punch force needs to supply the various types of work required in deep drawing, such as the ideal work of deformation, redundant work, friction work, and, when present, the work required for ironing. In this section, expressions for the force will be divided between the first drawing operation and the following drawing operations.

6.3.1 First Drawing Operation

In calculating the punch force for the first drawing operation, the radial stresses are sufficient for pure plastic deformation (neglecting friction) and are given by the following formula:

$$\sigma_p = 1.1k \ln \frac{D}{d_{s1}} \qquad (6.5)$$

where

D = diameter of blank

d_{s1} = mean diameter of cup after the first drawing

k = true stress

d_1 = inside diameter of cup after the first drawing operation

The theoretical force for pure plastic deformation is given by the formula:

$$F_p = \sigma_p \qquad (6.6)$$

Because of the many variables involved in this operation, such as friction, the blank holder force, and the die corner radius, all of which need to be included, the force for the first drawing in final form is given as:

$$F_1 = A_1 \sigma_1 = \pi d_{s1} T \sigma_1 \qquad (6.7)$$

where

$$\sigma_1 = e^{\mu \frac{\pi}{2}} \left(1.1k \ln \frac{D}{d_{s1}} + \frac{2\mu F_{d1}}{\pi d_{s1} T} \right) + k \frac{T}{2R_d + T}$$

$A1$ = cross–section area of cup after the first drawing

μ = coefficient of friction (Table 6.3)

k = main value of the true stress after the first drawing operation

$$k = \frac{k_0 + k_1}{2}; \quad k_0 = k(0) \text{ for } \varepsilon = 0$$

$$k_1 = k(\varepsilon_1) \text{ for } \varepsilon_1 = \frac{D - d_1}{D} \text{ or } k_1 = k(\varphi_1) \text{ for } \varphi = \ln \frac{D}{d_{s1}}$$

d_{s1} = mean diameter of cup after the first drawing

F_d = blank holder force

T = material thickness

R_d = die corner radius

D = blank diameter

F_d = blank holder force

T = material thickness

R_d = die corner radius

D = blank diameter

Table 6.3 Coefficient of friction μ

LUBRICANT	Drawing Material		
	Steel	Aluminum	Duralumin
Mineral oil	0.14 to 0.16	0.15	0.16
Vegetable oil	–	0.10	–
Graphite grease	0.06 to 0.10	0.10	0.08 to 0.10
No lubricant	0.18 to 0.20	0.35	0.22

6.3.2 Subsequent Drawing Operations

Subsequent drawing operations are different from the first drawing operation. In the latter, as the deep drawing proceeds, the flange diameter decreases. However, in the subsequent drawing operations, the zone of the plastic deformation does not change, so it is a steady–state process. The punch force for the next drawing operation can be calculated from the stress, which is given by the formula:

$$\sigma_i = 1.1 \left(1 + \frac{\mu\pi\alpha}{180}\right) \cdot (1.21 \text{ to } 1.44) k \left(1 + \frac{tg\alpha}{\mu}\right) \cdot \left[1 - \left(\frac{d_{si}}{d_{s(i-1)}}\right)^{\frac{\mu}{tg\alpha}}\right]$$

where

α = central conic angle of the drawing ring

d_{si}, $d_{s(i-1)}$ = the mean diameter i and $i - 1$ drawing operations

The punch force for the following operations is given by the formula:

$$F_i = A_i\sigma = \pi d_i T\sigma_i \tag{6.8}$$

Although these expressions for the punch force, when calculated, give much more precise results, sometimes a simple and very approximate formula is used to estimate the punch force, as follows:

$$F_i = \pi \cdot d_i \cdot T \cdot (UTS)\left(\frac{D}{d_i} - 0.7\right) \tag{6.8a}$$

where

d_i = punch diameter

D = blank diameter

T = material thickness

The equation does not include friction, the punch and die corner radii, or the blank holder force. However, this empirical formula makes rough provision for these factors. It has been established that the punch corner radius and the die radius, if they are greater than 10 times the material thickness, do not affect the maximum punch force significantly.

6.4 BLANK CALCULATIONS FOR ROTATIONAL SYMMETRICAL SHELLS

The volume of the developed blank before drawing should be the same as the volume of the shell after drawing.

$$V = V_{wp}$$

There are more methods for blank diameter calculation, such as the method of partial areas, the analytical method, the graphic method, and others.

6.4.1 Method of Partial Areas

Provided that the thickness of the material remains unchanged, the area of the workpiece will not change: thus, the blank diameter may be found from the area of the blank before drawing and the area of the shell after drawing.

$$A = A_{wp}$$

where

A = area of blank

A_{wp} = area of workpiece

The shell in Fig. 6.4a may be broken into matching components, illustrated in Fig. 6.4b, and the area of each component $A_1, A_2, A_3, \ldots A_n$ may be calculated. The area of the shell is the sum of the area components, found in the following equation:

$$A_{wp} = \sum_{i=1}^{i=n} A_i = A_1 + A_2 + A_3 + \ldots + A_n \qquad (6.9)$$

The diameter of the developed blank is

$$D = 0.13\sqrt{A_{wp}} \qquad (6.10)$$

where

D = diameter of the developed blank

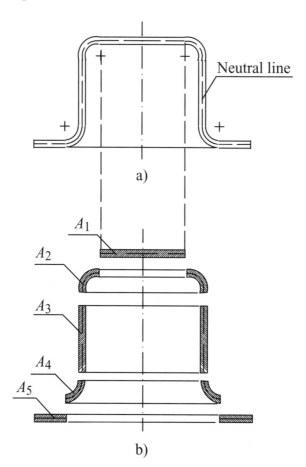

Fig. 6.4 Blank calculation for symmetrical shell: a) final workpiece; b) matching component of shell.

If the clearance c is large, the drawn shell will have thicker walls at the rim than at the bottom. The reason is that the rim consists of material from the outer diameter of the blank, which was reduced in diameter more than the rest of the shell wall. As a result, the shell will not have uniform wall thickness. If the thickness of the material as it enters the die is greater than the clearance between the punch and the die, then the thickness will be reduced by the process known as ironing.

The blank diameter for a drawn shell whose wall thickness is reduced may be found from the constant volume of the blank before drawing and the volume of the shell after drawing, with added correction for trimming of the drawn shell.

The blank volume is

$$V = V_{wp} + e_t V_{wp} = \pi \frac{D^2}{4} T$$

The diameter of the developed blank is

$$D = \sqrt{\frac{4V}{\pi T}} = 1.13 \left(\sqrt{\frac{V}{T}} \right) \qquad (6.11)$$

where

$\quad D$ = blank diameter

$\quad T$ = material thickness of blank

$\quad V$ = blank volume

$\quad e_t$ = percentage added for trimming

The value of e_t is given in Table 6.4.

Table 6.4 Percent value added for trimming (e_t)

Inner height/inner diameter of piece (h/d)	< 3	3 to 10	>10
$e_t\%$	8 to 10	10 to 12	12 to 13

6.4.2 Analytical Method

If the contour of a symmetrical shell is generated by the revolution of curve $f = f(x)$ in the xz-plane (with $x > 0$) rotated about the x axis for $2\pi = 360^0$, that surface of the shell can be calculated by using Guldinu's theorem, also known as Pappus's centroid theorem. The centroid is also called the center of mass, assuming the elementary curve has uniform line density. If the shell (Fig. 6.5) is divided into more elements with known length $l_1, l_2, l_3...l_i...l_n$, and its distance of center of mass is $r_1, r_2, r_3 ... r_i ... r_n$, then the surface of the workpiece can be calculated by the following equation:

$$A_{wp} = 2\pi \sum_{i=1}^{n} l_i r_i$$

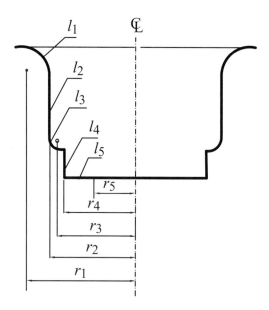

Fig. 6.5 Method for determining the diameter of a noncylindrical shell.

The centroid of straight lines corresponding with the middle of the lines, and the centroid of the arcs of the circle can be calculated by the formulas given in Table 6.5.

Table 6.5 Centroid position of arc of circle

D I A G R A M	a)	b)	c)
α^0	$\eta = \dfrac{r\sin(\alpha/2)\cdot 180^0}{\pi\alpha}$	$x = \eta\sin\dfrac{\alpha}{2}$	$x = \eta\cos\dfrac{\alpha}{2}$
30°	0.988r	0.256r	0.955r
45°	0.978r	0.373r	0.901r
60°	0.955r	0.478r	0.727r
90°	0.900r	0.637r	0.637r

The area of the developed blank A before drawing should be the same as the area of the shell A_{wp} after drawing. Therefore,

$$A = A_{wp} = \frac{\pi D^2}{4} = 2\pi \sum_{i=1}^{i=n} l_i r_i \tag{6.12}$$

and, hence, the diameter of the developed blank D is

$$D = \sqrt{8 \sum_{i=1}^{i=n} l_i r_i} = 2.83 \left(\sqrt{\sum_{i=1}^{i=n} l_i r_i} \right) \tag{6.13}$$

6.4.3 Influence of Wall Thickness on Blank Calculations

All of the previous studies are subject to the proviso that the thickness of the material remains unchanged in all cross–sections of a drawn workpiece. Because the punch exerts a force on the cup bottom to cause the drawing action, and a downward holding force is applied by the blankholder, considerable stretching of the metal occurs in the cup's side walls. However, compressive forces involved on the outer edge of the blank tend to thicken the metal.

The thinning and thickening of metal in the cup drawing operation also indicates the flow of the metal. The normal variation of wall thickness in cup drawing is illustrated in Fig. 6.6.

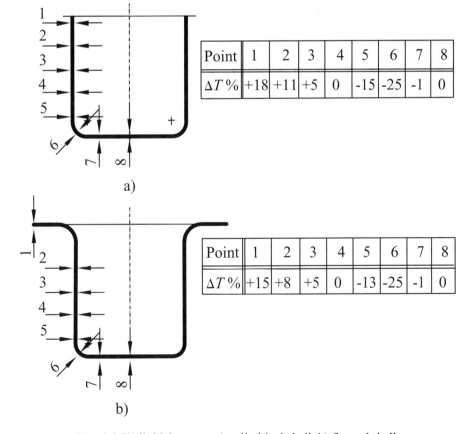

Point	1	2	3	4	5	6	7	8
$\Delta T\%$	+18	+11	+5	0	-15	-25	-1	0

a)

Point	1	2	3	4	5	6	7	8
$\Delta T\%$	+15	+8	+5	0	-13	-25	-1	0

b)

Fig. 6.6 Wall thicknesses: a) cylindrical shell; b) flanged shell.

Note that the maximum point of the wall thinning occurs in a cup with a spherical bottom. On flat–bottom cups, no change in thickness of material occurs at the cup bottom because the stress here is too low to cause permanent metal deformation. Maximum wall thickening occurs on the outer edge of the drawn cup.

The most severe thinning occurs near the bottom at the zone of the bottom radius where there is maximum tensile stress. In extreme cases, a material can be up to 35 percent thinner than the starting thickness of blank in this zone.

The maximum wall thickness on cylindrical shell without flange can be calculated by the following formula:

$$T' = T\left(\sqrt{\frac{D}{d}}\right) \tag{6.14}$$

The maximum thickness of the material in a shell with flange occurs on the outer edge of the flange, and can be calculated by the following formula:

$$T' = T\left(\sqrt{\frac{D}{d_f}}\right) \tag{6.15}$$

where

   T' = maximum thickness of the drawn shell

   T = thickness of blank

   D = blank diameter

   d = diameter of shell

   d_f = diameter of flange.

The median material thickness of a drawn shell can be calculated by the following formula:

$$T'_{med} = \frac{T'_1 + T'_2 + \dots + T'_i + \dots + T'_n}{n} \tag{6.16}$$

The volume of the developed blank should be the same as the volume of the shell after drawing, that is:

$$A \cdot T = A_{wp} \cdot T_{med} \quad \text{or} \quad \frac{\pi D^2}{4} = A_{wp} \cdot \frac{T_{med}}{T} = A_{wp} \cdot k$$

Hence, the blank diameter is

$$D = \sqrt{k}\left(\sqrt{\frac{4}{\pi}\sum_{i=1}^{n} A_i}\right) = \sqrt{k}\left(\sqrt{8\sum_{i=1}^{n} l_i r_i}\right) \tag{6.17}$$

where

   k = factor correction

Factor correction k is dependent on the relative radius R and the specific pressure of the blankholder p_h.

$$R = \frac{R_k + R_i}{T}$$

where

   R_k = die ring radius

   R_i = punch radius

   T = material thickness

Factor correction $k = f(R, p_h)$ is given in Table 6.6.

Table 6.6 Factor correction (k)

$k = f(R, R\, p_h)$	Specific blankholder pressure (p_h) (N/mm²)	Relative radius $R = \dfrac{R_k + R_i}{T}$
1.00 – 0.969	1.0 – 2.0	> 3
0.97 – 0.929	2.1 – 2.5	3 – 2
0.93 – 0.900	2.51 – 3	< 2

6.4.4 Allowance for Trimming

When sheet metal is rolled, a fiber structure is formed in the direction of rolling. The sheet metal is stronger and has greater allowed elongation in the direction of rolling. Strength is lower across the direction of rolling. Due to this variation in strength, the top edge of the shell is wavy. This wavy phenomenon is called *earing*. This objectionable wavy edge increases the height of a drawn cap and must be allowed for (as allowance for trimming) when determining the blank diameter.

The height of the drawn shell h that is used to calculate the blank diameter is

$$h = h_c + \Delta h$$

where

h = height of drawn shell

h_c = height of final shell

Δh = trim allowance

The values of Δh are given in Table 6.7. These values are approximate and should be used only for trial calculation. The exact trim allowance can be determined more exactly by practice and experiment.

Table 6.7 The trim allowance Δh for a cylindrical cup without a flange

h_c (mm)	Ratio d/h_c			
	0.50 to 0.80	0.81 to 1.60	0.161 to 2.50	2.51 to 4.00
10	1.5	1.7	2.0	2.6
20	1.7	2.0	2.5	3.2
50	2.5	3.0	3.6	4.5
100	3.5	4.2	5.8	6.8
150	4.5	5.5	6.5	8.3
200	5.8	6.5	8.5	10.8
250	7.0	7.8	9.8	11.5
300	8.0	9.5	11.0	13.0

A flange trim allowance is added to the flange diameter in a cylindrical cup, so the diameter of the flange before trimming is

$$d_1 = d_f + \Delta d$$

where

   d_1 = diameter of flange before trimming

   d_f = diameter of final flange

   Δd = trimming allowance

   The trim allowance for cups with flange Δd depends on the diameter of the cup d and the ratio of the flange diameter d_f and the diameter of the cup d_f/d. The value of Δd is given in Table 6.8.

Table 6.8 The trim allowance Δd for a cylindrical cup with flange

d (mm)	Ratio d_f/d			
	to 1.5	1.51 to 2.0	2.1 to 2.50	2.51 to 3.00
25	1.8	1.5	1.7	1.2
50	2.5	2.3	2.0	1.8
100	3.5	3.2	2.6	2.3
150	4.5	3.8	4.2	3.5
200	5.3	4.5	3.9	3.7
250	6.2	5.0	4.2	3.9
300	6.8	5.7	4.6	4.0

   Because the trim allowance used is directly related to the percentage of scrap produced, care must be taken in setting this allowance. More trim allowance is made in progressive drawing dies. In these dies, the cup is drawn slightly off center from the blank, due to variations in the location of the striper skeleton.

Example:

Calculate the blank diameter for the piece shown in Fig. 6.7.

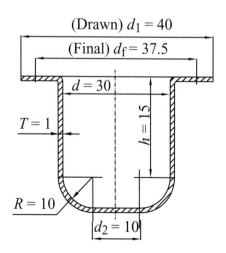

Fig. 6.7 Final drawn cylindrical cup with flange.

Solution:

The final cup has the diameter of the flange $d_f = 37.5$ mm. The diameter of the drawn cup before trimming is:

$$d_1 = d_f + \Delta d$$

The trimming allowance from Table 6.8 is

$$\frac{d_f}{d} = \frac{37.5}{30} = 1.25 < 1.5 \text{, and } d_f = 37.5 \text{ mm is } \Delta d = 2.5 \text{ mm}$$

Hence, the diameter of the flange before trimming is

$$d_1 = 37.5 + 2.5 = 40 \text{ mm}$$

In the following blank diameter calculations, diameter d_1 is used:

a) Method of partial surfaces (Fig. 6.8)

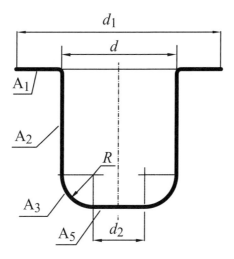

Fig. 6.8 Method of partial surfaces.

$$A_1 = \frac{\pi}{4}\left(d_1^2 - d^2\right) = 0.785\left(40^2 - 30^2\right) = 549.5 \text{ mm}^2$$

$$A_2 = \pi \cdot d \cdot h = 3.14 \cdot 30 \cdot 15 = 1413 \text{ mm}^2$$

$$A_3 = \frac{\pi}{4}\left(2\pi R d_2 + 8R^{2)}\right) = 0.785(6.28 \cdot 10 \cdot 10 + 8 \cdot 10^2) = 1121 \text{ mm}^2$$

$$A_4 = \frac{\pi}{4} d_2^2 = 0.785 \cdot 100 = 78.5 \text{ mm}^2$$

$$A_{wp} = \sum_{i=1}^{n} A_i = A_1 + A_2 + A_3 + A_4 = 549.5 + 1413 + 1121 + 78.5 = 3162 \text{ mm}^2$$

Blank diameter (Equation 6.10):

$$D = 1.13\left(\sqrt{A_{wp}}\right) = 1.13\left(\sqrt{3162}\right) = 63.5 \text{ mm}$$

b) Analytical method (Fig. 6.9)

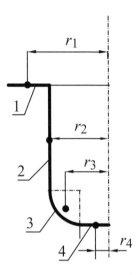

Fig. 6.9 Analytical methods.

1. Calculation of length of line:

$$l_1 = \frac{d_1 - d}{2} = \frac{40 - 30}{2} = 5.0 \text{ mm}$$

$$l_2 = h = 15 \text{ mm}$$

$$l_3 \frac{\pi R}{2} = \frac{\pi \cdot 10}{2} = 15.7 \text{ mm}$$

$$l_4 = \frac{d_2}{2} = \frac{10}{2} = 5 \text{ mm}$$

2. Calculation distance of centroid:

$$r_1 = \frac{d}{2} + \left(\frac{d_1 - d}{4}\right) = \frac{30}{3} + \frac{40 - 30}{4} = 17.5 \text{ mm}$$

$$r_2 = \frac{d}{2} = \frac{30}{2} = 15 \text{ mm}$$

$$r_3 = \frac{d_2}{2} + 0.637R = \frac{10}{2} + 0.637 \cdot 10 = 11.37 \text{ mm (Table 6.5 for arc centroid definition)}$$

$$r_4 = \frac{d_2}{4} = \frac{10}{4} = 2.5 \text{ mm}$$

Blank diameter (Equation 6.13):

$$D = \sqrt{8\sum_{i=1}^{i=n} l_i r_i} = 2.83\left(\sqrt{\sum_{i=1}^{4} l_i r_i}\right) = 2.83\left(\sqrt{l_1 r_1 + l_2 r_2 + l_3 r_3 + l_4 r_4}\right) = 2.83\left(\sqrt{87.5 + 225 + 178.5 + 12.5}\right)$$

$$D = 2.83\left(\sqrt{503.5}\right) = 63.5\,\text{mm}$$

c)   Using equation for calculation blank diameter (Appendix 1 equation # 24):

$$D = \sqrt{d_1^2 + 4d9h + 0.57R) - 0.57R^2} = \sqrt{40^2 + 4\cdot 30(15 + 0.57\cdot 10) - 0.57\cdot 10^2}$$

$$D = 63.5 \text{ mm}.$$

6.4.5 Shells with Reduced Thickness of Wall

For calculating the blank diameter when the shell needs to be drawn with reduction of wall thickness (Fig. 6. 10), the volume of the mass of the blank before drawing should be the same as the volume of the mass of the shell after drawing, but before trimming. The blank volume with trim allowance is:

$$V = \frac{\pi D^2}{4}T = (1+e)V_{wp}$$

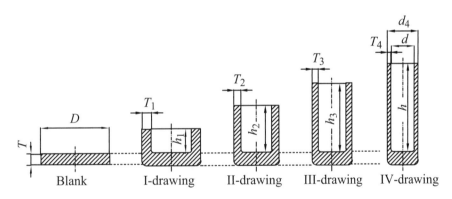

Fig. 6.10 Deep drawing with reduction of wall thickness.

The blank mass with trim allowance is:

$$m = \gamma \cdot V = \gamma \frac{\pi D^2}{4}T = (1+e)m_{wp}$$

The blank diameter should be then calculated as follows:

$$D = \sqrt{\frac{4V}{\pi T}} = 1.13\left(\sqrt{\frac{V}{T}}\right) = 1.13\left(\sqrt{\frac{m}{\gamma T}}\right) \tag{6.18}$$

where

D = blank diameter

T = blank thickness

V = blank volume $\left[V = \left(1 + e\right)V_{wp} \right]$

m = blank mass $\left[m = \left(1 + e\right)m_{wp} \right]$

V_{wp} = volume of drawn shell

e_t = trim allowance

γ = specific mass

The trim allowance value depends on the ratio of the inner shell height h and the inner diameter d of the shell. These values are given in Table 6.9.

Table 6.9 Trim allowance e for drawn shell with wall thickness reduction

Ratio h/d	< 3	3 to 10	> 10
e_t %	8 to 10	10.1 to 12.0	12.1 to 15.0

Example:

Define the blank dimension for drawing a shell with a reduction of wall thickness (Fig. 6.10). Assume the following:

Outer diameter of shell = 30 mm

Wall thickness = 0.8 mm

Height h = 115 mm

Bottom thickness T = 4 mm

Inner diameter $d = d_4 - 2T_4 = 30 - 2 \cdot 0.8 = 28.4 \, \text{mm}$

Solution:

The blank thickness is the same as the thickness of the bottom $T = 4$ mm. The volume of the blank is

$$V = (1 + e)V_{wk}$$

For the ratio $\dfrac{h}{d} = \dfrac{115}{28.4} = 4.04$, the trim allowance, according to Table 6.9, is $e = (10.1 \text{ to } 12)\%$.

The volume of the shell is

$$V_{wp} = V_b + V_w = \frac{\pi d_4^2}{4}T + \frac{\pi \left(d_4^2 - d^2\right)}{4}h = \frac{\pi 30^2}{4}4 + \frac{\pi \left(30^2 - 28.4^2\right)}{4}115 = 11,261 \text{ mm}^3$$

The volume of the blank is

$$V = \left(1 + e\right)V_{wp} = 1.12 \cdot 11,261 = 12,612.3 \text{ mm}^3$$

The diameter of the blank is (Equation 6.13)

$$D = 1.13\left(\sqrt{\frac{V}{T}}\right) = 1.13\left(\sqrt{\frac{12,612.3}{4}}\right) = 56.12 \text{ mm}$$

The dimensions of the blank are diameter $D = 56.15$ mm and thickness $T = 4$ mm.

6.5 DETERMINING SHAPE OF BLANK FOR NONSYMMETRICAL SHELLS

6.5.1 Rectangular Shells

The corner radius R_e of a rectangular– or square–shaped shell with dimensions a, b, and c in Fig. 6.11 is the major limiting factor as to how deeply this shell can be drawn in one stroke of the press. There are other factors that also have an influence on the maximum depth, such as the relation between the bottom radius R_b and the corner radius R_e, the minimum length between the corner radii, and the drawing speed, etc.

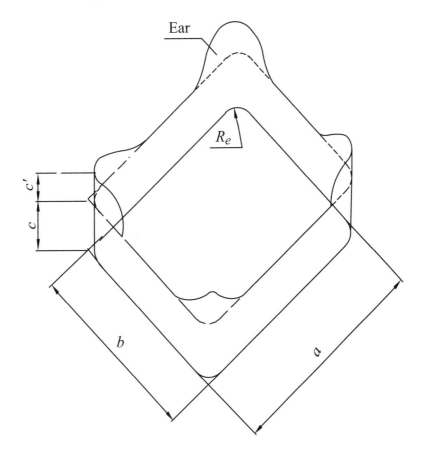

Fig. 6.11 Earing in a drawn rectangular shell.

There is no formula for determining the shape of the blank for rectangular drawing that will produce the part as drawn to print. All corner contours must be developed. However, the following will get the die in the ballpark safely and with a minimum of trials

There are three possibilities for developing a corner contour and all are dependent on the relationship of R_b to R_e.

1. If $R_b < R_e$, the blank radius R of the corner contours in Fig 6.12 is given by the formula:

$$R = \sqrt{R_e^2 + 2R_e\left(c - 0.47R_b\right)} \tag{6.19}$$

where

$c = c_0 + \Delta c$-height of drawn workpiece

Δc = material added for trimming

c_0 = height of final part

Some values of Δc are given in Table 6.10

Table 6.10 Values of material added for trimming (Δc)

c_0 / R_e	Number of draws	Δc (mm)
2.5 to 7.0	1	$(0.03 \text{ to } 0.05) \cdot c_0$
7.0 to 18.0	2	$(0.04 \text{ to } 0.06) \cdot c_0$
18.0 to 45.0	3	$(0.05 \text{ to } 0.08) \cdot c_0$
45.0 to 100	4	$(0.06 \text{ to } 0.10) \cdot c_0$

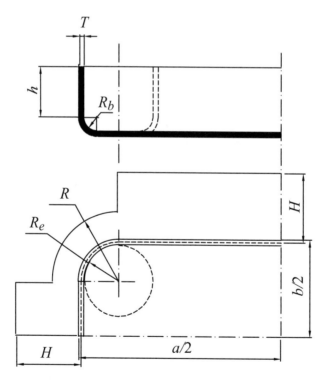

Fig. 6.12 Developing corner contour: $R_b < R_e$.

2. If $R_b = R_e$, the blank radius R of the corner contours in Fig 6.13 is given by the formula:

$$R = \sqrt{2cR_e} \tag{6.20}$$

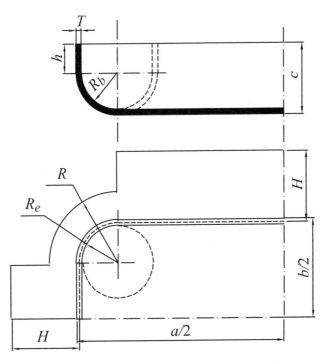

Fig. 6.13 Developing corner contour: $R_b = R_e$.

3. If $R_b = 0$, the blank radius R of the corner contours in Fig 6.14 is given by the formula:

$$R = \sqrt{R_e^2 + 2cR_e} \tag{6.21}$$

In all three examples, the center of the radii R and R_e is the same. Considering that the flat shell sides are bent, for calculating the flat blank dimensions of H, the following formula may be used:

$$H = h + 0.57R_b \tag{6.22}$$

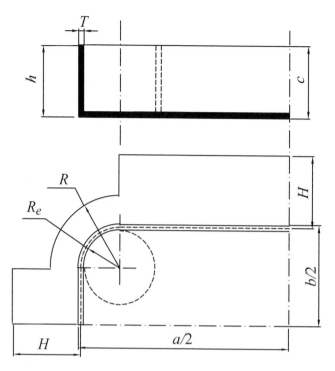

Fig. 6.14 Developing corner contour: $R_b = 0$.

However, calculating the flat blank size for rectangular drawn shells (Fig. 6.12, Fig. 6.13, and Fig. 6.14) in this way is not satisfactory because the sharp transition between the corner arcs and the flat sides will result in cracks. The shape of the blank needs to be modified as shown in Fig. 6.15, according to the following steps:

a) Draw a rectangle with dimensions a and b.

b) At each side of the rectangle, add a value H.

c) From the center of radius R_e (point O), draw an arc with radius $R_1 = xR$, where

$$x = 0.0185 \left(\frac{R}{R_e} \right)^2 + 0.982$$

d) Reduce the height of each side by the following values:

$$H_{sa} = 0.785 \left(x^2 - 1 \right) \cdot \frac{R^2}{a} \qquad (6.23)$$

$$H_{sb} = 0.785 \left(x^2 - 1 \right) \cdot \frac{R^2}{b}$$

e) Round the corners by radii R_a and R_b, whose value is defined graphically. Note that the subtracted surfaces should be equal to the added surfaces.

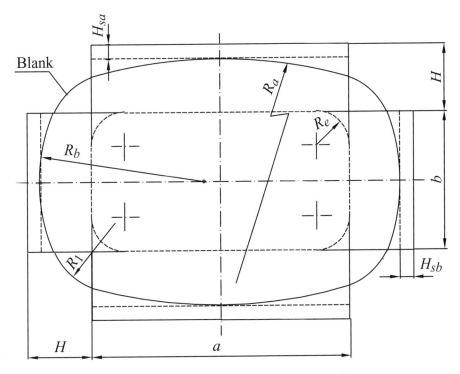

Fig. 6.15 Modified blank of rectangular shell.

When laying out the blank it is usually advisable to plan for a form that will produce corners a little higher than the sides. The wear on the die is at the corners, and when it occurs, the metal will thicken and the drawn part will be low at the corners if no allowance for wear has been made on the blank.

6.5.2 Square Shells

A flat blank for square shells without flanges (Fig. 6.16) has a circular shape whose diameter may be calculated by the formula

$$D = 1.13\sqrt{a^2 + 4a(c - R_b) + 2.28R_e^2 - R_e(1.72c - 5.3R_b)} \tag{6.24}$$

The height of the workpiece is $c = c_0 + \Delta c$; the value of Δc may be found from Table 6.10, or calculated by the formula: $\Delta c = (0.7 \text{ to } 0.8)\sqrt{c}$.

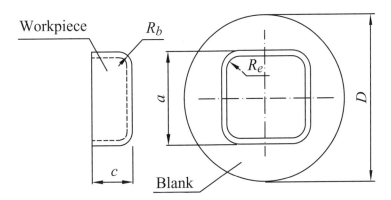

Fig. 6.16 Blank for square shell.

6.6 DRAWING PRACTICE

6.6.1 Defects During Deep Drawing

Deep drawing operations are governed by many complex factors that may result in either successful or defective products.

Clearance. If the shell fractures during the deep drawing operation, the problem may be that the clearance between the punch and the die is incorrect. This problem can be a direct result of the punch and die having been designed or made with incorrect clearance. Chapter 11 may help in making a correct choice for the clearance. Fractures can also result if the thickness of the work piece is out of tolerance or not uniform, or if the punch and die are not properly aligned.

Blank holder pressure. If too much force is applied to the blank, the punch load will be increased (because of the increase in friction), and this increase will lead to fracture of the shell wall. To calculate the blank holder force, see Chapter 11.

The corner radius of the punch and die radius. These radii are important for a successful deep drawing operation. If the radii are too small, the corner may fracture because of the increased force required to draw the cup. Scratches, dirt, or any surface defect of the punch or die increase the required drawing force and may cause a shell to tear (Fig. 6.3a and Fig. 6.3b). For correct calculation of the corner punch and die radii, see Chapter 11. If the blank holder exerts too little pressure, or if the die radius is too large, wrinkles will appear at the top flange of the part, as shown in Fig 6.17.

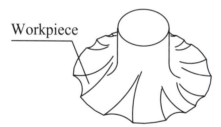

Fig. 6.17 Wrinkles on the workpiece.

Drawing beads are useful in controlling the flow of the blank into the die cavity. They are necessary for drawing nonsymmetrical shells. For the proper design and location of drawing beads, see Chapter 11.

6.6.2 Lubrication in Drawing

During deep drawing, different lubrication conditions exist, from hydrodynamic lubrication in the blank holder to boundary lubrication at the drawing radius, where breakdown of the film very often occurs. Lubrication in deep drawing is important in lowering forces, increasing drawability, reducing wear of the tool, and reducing defects in the workpiece. Lubricant selection is based on the difficulty of the operation, the type of drawing operation; and the material; recommendations are given in Table 6.11. In this table, a *mild* operation typically is a shallow draw on low–carbon steel, a *medium* operation is a deep draw on low–carbon steel, and a *severe* operation is a cartridge–case draw or a seamless tube draw.

Table 6.11 Lubricants commonly used in deep drawing processes

DRAWING OF MATERIAL	LUBRICANT
Steel – Carbon and Low Alloy	*Mild Operation:* Mineral oil of medium heavy–to–heavy viscosity, Soap solutions (0.03 to 2.0 percent high–titer soap), Fatty oil + mineral oil, emulsions, and lanolin. *Medium Operation:* Fat of oil in soap–base emulsions, fatty oil + mineral oil, soap + wax, dried soap film. *Severe Operation:* Dried soap or wax film, sulphide or phosphate coatings + emulsions with finely divided fillers and sometimes sulphurized oils.
Stainless Steel	*Mild Operation:* Corn oil or castor oil, castor oil + emulsified soap, waxed or oiled paper. *Medium Operation:* Powdered graphite suspension dried on work piece before operation (to be removed before annealing), solid wax film. *Severe Operation:* Lithopone and boiled linseed oil, white lead and linseed oil ina heavy consistency.
Aluminum and Aluminum–Alloys	*Mild Operation:* Mineral oil, fatty oil blends in mineral oil (10 to 20 % fatty oil). Tallow and paraffin, sulphurized fatty oil (blends 10 to 15%) preferably enriched with 10% fatty oil. *Severe Operation:* Dried soap film or wax film, mineral oil or fatty oil, fat emulsions in soap water + finely divided fillers.
Titanium	Chlorinated paraffin, soap, polymer, and wax.
Copper	Fatty oil + soap emulsions, fatty oil + mineral oil, lard oil blends (25 to 50%) in mineral oil, dried soap properly applied.

Seven

Various Forming Processes

7.1 STRETCH FORMING

The process of stretch drawing was developed as a method of putting metals under combined bending and tension stresses at the same time. Sometimes, a part that has been previously bent may be used as an initial material in stretch draw forming. In stretch forming, the sheet is clamped around its edges and stretched over a die or form block. This process strains the metal beyond the elastic limit to set the work-piece shape permanently.

Workpieces may have single or double curvatures, as in aircraft skin panels and structure frames, or automobile body parts. To assess the formability of sheet metals while forming a workpiece, circle grid analysis is used to construct a forming limit diagram of the material to be used. In such an analysis, a circular pattern is etched on the sheet blank The blank is then formed in a die. Each circle on the blank will deform in a different manner due to local forming patterns. After a series of such tests on a particular metal sheet, the deformed circles are analyzed to produce a forming limit diagram (FLD) that shows the overall forming pattern of the blank during plastic deformation.

In the forming limit diagram, the major strain is always positive. However, minor strains can be positive and negative at the same time. Fig. 7.1 shows the FLD that bounds the deformation of the sheet metal. Above the curves is the failure zone, and below the curves is the safe zone; the actual strains used in stretch draw forming must be below the curve for any given material.

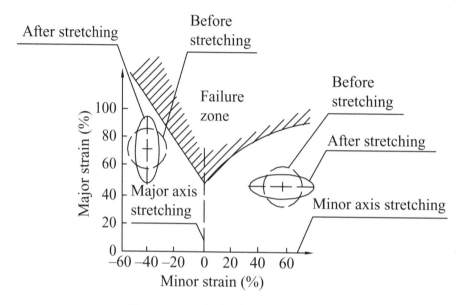

Fig. 7.1 Forming limit diagram (FLD).

Two methods are used in stretch forming: the form block method and the mating die method.

a) Form Block Method

The form block method is shown in Fig. 7.2. Each end of the blank is securely held in tension by an adjustable gripper, which is moved to stretch the blank over a form block. The desired shape of the workpiece is formed by the action of the form block as the material is moved hydraulically against the block.

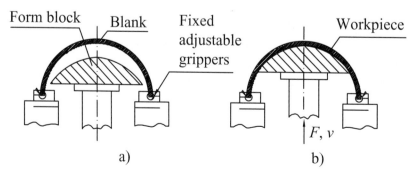

Fig. 7.2 Stretch draw forming with a form block: a) starting position; b) final position.

b) Mating Die Method

The mating die method is shown in Fig. 7.3. The blank is held in tension by grippers which, as they move, perform two actions. First, they stretch the workpiece by a predetermined amount to approximately 2% elongation over the form block. The punch then descends onto the blank to form the workpiece into the desired shape by pressing the metal against the dies. The process is used primarily for aerospace and automobile applications with a variety of materials. The workpieces may have single or double curvatures, such as in aircraft wings or fuselage skin panels, and automobile body parts. Typical shapes of workpieces formed by these methods are shown in Fig. 7.4.

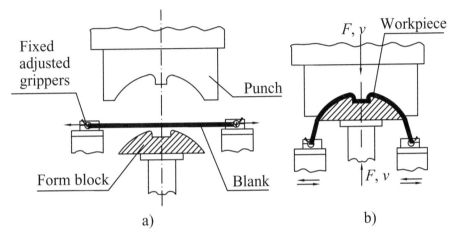

Fig. 7.3 Stretch draw forming with mating dies: a) blank is held in tension; b) punch moves to form workpiece.

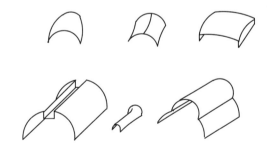

Fig. 7.4 Typical shapes of part formed with mating dies.

7.2 NOSING

Nosing is a die reduction method whereby the top of a cup or a tubular shape may be made closer or smaller in diameter than its body.

There are three types of end profiles after nosing:

- frustum of cone
- neck
- segment of sphere

If the material is not too thin, it is possible to reduce the top of the cup to about 20 % of its diameter in one operation. Nosing compresses the work metal, resulting in an increase in length and wall thickness. Fig. 7.5 shows schematic illustrations of three types of die reduction methods. The calculation of the height of a drawn shell or tubing by nosing its end is different for each type.

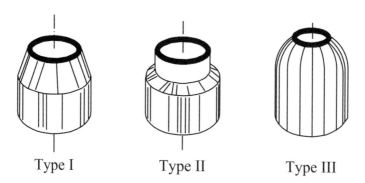

| Type I | Type II | Type III |

Fig. 7.5 Schematic illustration of different types of nosing.

a) Type I Reductions

Fig. 7.6 shows a nosing cup in the shape of a frustum of a cone. The height H of a shell before the nosing operation may be calculated by the following formula:

$$h = 1.05(h_1) + \frac{D^2 - d^2}{4\sqrt{Dd}\sin\alpha} \tag{7.1}$$

where

D = mean diameter of a drawn cup before nosing

d = main diameter of a workpiece after nosing

α = angle of cone

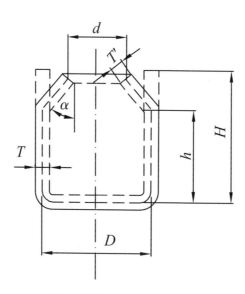

Fig. 7.6 Nosing shell type I

Force – The nosing force for Type I reductions may be calculated by the formula below:

$$F_\mathrm{I} = \pi D T k_{\mathrm{sr}} \left(1 - \frac{d}{D}\right) C_1 C_2 \tag{7.2}$$

where

$$k_{\mathrm{sr}} = 0.5 \left(k_0 + k'\right)$$

$$C_1 = 0.5 \left(1 + \sqrt{\frac{D}{d}}\right); \quad C_2 = \left(1 + \mu \cdot \mathrm{ctg}\,\alpha\right)\left(3 - 2\cos\alpha\right)$$

$k_0 = k\,(0)$ for $\varepsilon = 0$ — specific deformation impedance non-nosing part of shell;

$k' = k\left(\varepsilon'\right)$ for $\varepsilon' = 1 - \dfrac{d}{D}$ — specific deformation impedance nosing part of shell T = material thickness

b) Type II Reductions

For a necking type of nosing, as shown in Fig. 7.7, the height of the workpiece before necking may be calculated by the formula:

$$H = 1.05 \left(h_1 + \frac{h_2 d \dfrac{D^2 - d^2}{4\sin\alpha}}{\sqrt{D \cdot d}}\right) \tag{7.3}$$

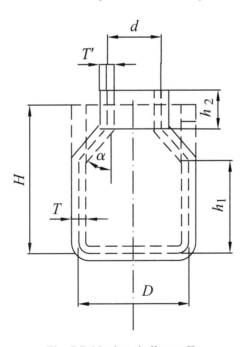

Fig. 7.7 Nosing shell type II.

Force – The necking force for Type II reductions is given by the following formula:

$$F_\mathrm{II} = F_\mathrm{I} + F_\mathrm{s} \tag{7.4}$$

where

$$F_1 = \text{from equation (7.2)}$$

$$F_s = 1.82 \cdot k \frac{T'^2}{R_m} C_3$$

$$C_3 = d + R_m \left(1 - \cos \alpha\right)$$

T' = material thickness of necked part of shell

$$R_m = \text{die ring corner radius}$$

c) Type III Reductions

For a segment of sphere type nosing, as shown in Fig. 7.8, the height of the shell before the nosing operation may be calculated by the following formula:

$$H = h + 0.25 \left(1 + \sqrt{\frac{D}{d}}\right)\sqrt{D^2 - d^2} \tag{7.5}$$

Note: This equation is applied for $R_1 = 0.5D$.

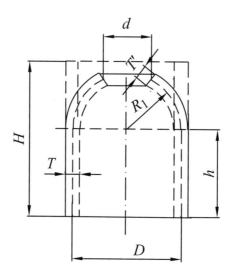

Fig.7.8 Nosing shell type III.

Force – The force for Type III reductions is given by the following:

$$F_{III} = \pi \cdot d \cdot T \cdot k_{sr} \left(1 - \frac{d}{D}\right) C_1 C_4 \tag{7.6}$$

where

$$C_1 = 0.5 \left(1 + \sqrt{\frac{D}{d}}\right)$$

$$C_4 = \mu \cdot \sqrt{\frac{4R_1}{D-d} + \frac{T}{4R_1}}$$

$$\mu = \text{coefficient of friction}$$

The thickness of the material in the deformation zone of the workpiece becomes greater as a result of nosing. The maximum thickness (T') at the end of the workpiece, after the nosing operation, may be calculated by the following formula:

$$T' = T\left(\sqrt{\frac{D}{d}}\right) \tag{7.7}$$

The median nosing ratio is given by the formula:

$$m_s = \frac{d_1}{D} = \frac{d_2}{d_1} = \dots \frac{d_n}{d_{n-1}} \tag{7.8}$$

Values of m_s for steel and brass are given in Table 7.1.
For the first operation, the value of the nosing ratio is:

$$m_1 = 0.9m_s \tag{7.8a}$$

For subsequent operations, the value of the nosing ratio is 5 % to 10% greater than *ms* from Table 7.1. The nosing diameter for the final phase of nosing is given by this formula:

$$d_n = m_s^n D \tag{7.9}$$

From formula (7.9), the necessary number of nosing operations is:

$$n = \frac{\log d_n - \log D}{\log m_s} \tag{7.10}$$

Table 7.1 Median values of nosing ratio m_s for different materials

MATERIAL	MATERIAL THICKNESS *T* (mm)		
	< 0.5	0.5 to 1.0	> 1.0
Steel	0.80	0.75	0.70 to 0.65
Brass	0.85	0.80 to 0.70	0.70 to 0.65

7.3 EXPANDING

Expanding is a process that is used to enlarge the diameter of the drawn shell or tube in one or more sections by a different kind of punch, such as a flexible plug (rubber or polyurethane), by hydraulic pressure, or by a segmented mechanical die. Fig. 7.9 shows the characteristic shapes of shells formed by expanding methods.

Sizing is a term used to describe the further flattening or improvement of a selected surface on previously drawn parts to a closer limit of accuracy than is possible by conventional drawing methods. Sizing consists of squeezing the metal in a desired direction.

Bulging basically involves placing a tubular, conical, or drawn workpiece in a split female die and expanding it, such as by means of a flexible plug (rubber or polyurethane). The punch is then retracted and the part is removed by opening the die. The major advantage of using polyurethane plugs is that they are resistant to abrasion, wear, and lubricants. Hydraulic pressure can also be used in this operation, but will require sealing and hydraulic controls.

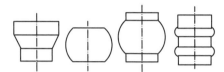

Fig.7.9 Characteristic shapes of shells formed by expanding.

Segmented tools are usually used for forming cone rings, sizing cylindrical rings, and expanding tubular and drawn shells. These tools do not have a female die; the workpiece pulls on the punch, which has the shape of the final part. Expansion is thus carried out by expanding the punch mechanically. Fig. 7.10 shows the action of force in segmented punches.

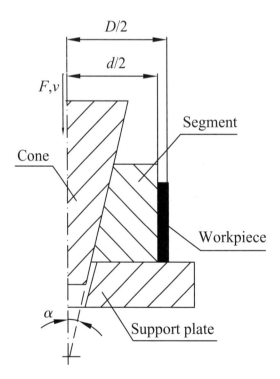

Fig. 7.10 Schematic illustration of segmented punch.

The expanding force for this type of tool is given by the formula:

$$F = \frac{1.1k\left(tg\alpha + \mu\right)\pi db}{1 - \left(\mu_1 + \mu_2\right)tg\alpha - \mu_1\mu_2}\ln\frac{D}{d}$$
(7.11)

where

α = half angle of cone

μ_1 = coefficient of friction between cone and segmented punch

μ_2 = coefficient of friction between segments and lower (supporting) plate of tool

D, d = outer and inner diameters of workpiece

k = specific deformation impedance

b = width of workpiece

A method of enlarging portions of a drawn shell in a press is shown in Fig. 7.11. The value of the expanding ratio $K = D/d$ is dependent on the kind and heat-treatment condition of the material and the relative thickness of the material d/T. Some values of K are given in Table 7.2. The thickness of the material at the deformation zone of the workpiece is reduced by the expansion operation and is given by the formula:

$$T' = T\left(\frac{d}{D}\right)$$
(7.12)

where

T = thickness of the material before expanding

T' = thickness of the material in the deformation zone

d, D = main diameter of workpiece before and after expanding

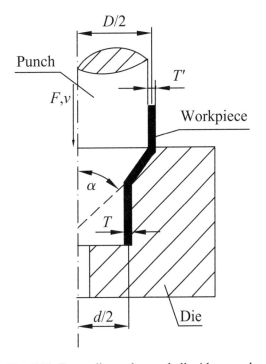

Fig. 7.11 Expanding a drawn shell with a punch.

Table 7.2 Values of expanding ratio K for enlarging portions of a drawn shell

MATERIAL	Relative thickness of material $(d/T) \times 100\ \%$			
	0.45 to 0.35		0.32 to 0.28	
	Annealed	Hard	Annealed	Hard
Low-carbon steel	1.20	1.10	1.15	1.05
Aluminum and copper	1.25	1.20	1.20	1.15

The punch force can be calculated by the formula:

$$F = \frac{\pi dTk}{3 - \mu - 2\cos\alpha}\left(\ln\frac{D}{d} + \sqrt{\frac{T}{D}}\sin\alpha\right) \tag{7.13}$$

where

$$\left.\begin{array}{l} k = 0.5 k_{\text{sr}}\left(k_0 + k_1\right) \\ k_0 = k(0) \ \text{ for } \ \varepsilon = 0 \\ k_1 = k(\varepsilon) \ \text{ for } \ \varepsilon = \ln\dfrac{D}{d} \end{array}\right\} = \text{main deformation impendence}$$

7.4 DIMPLING

Dimpling is the process of bending and stretching (flanging) the inner edges of sheet metal components (Fig. 7.12). A hole is drilled or punched and the surrounding metal is expanded into a flange. Sometimes a shaped punch pierces the sheet metal and is expanded into the hole. Stretching the metal around the hole subjects the edges to high tensile strains, which could lead to cracking and tearing. As the ratio of flange diameter to hole diameter increases, the strain increases proportionately. The diameter of the hole may be calculated by the formula:

$$d = D - \left(2H - 0.86R_{\text{m}} - 1.43T\right) \tag{7.14}$$

where

d = hole diameter before dimpling

H = height of flange

R_{m} = die corner radius (Table 7.3)

T = material thickness

Table 7.3 Values of die corner radius R_{m}, for dimpling

Material thickness (mm)	Die corner radius R_m (mm)
$T \leq 2$	(4 to 5)T
$T \geq 2$	(2 to 3)T

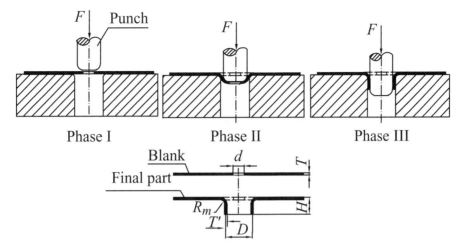

Fig. 7.12 Phases in the dimpling process.

The height of the flange is given by the formula:

$$H = \frac{D-d}{2} + 0.43R_{\mathrm{m}} + 0.715T \tag{7.15}$$

The punch force is given by:

$$F = 1.1\pi T\left(UTS\right)\cdot\left(D-d\right) \tag{7.16}$$

The thickness of the material at the end of the flange is reduced by the stretching of the material. The minimum thickness of cylindrical flanges is given by the formula:

$$T' = T\cdot\sqrt{\frac{d}{D}} \tag{7.17}$$

In the dimpling process, the ratio of the hole diameter to the flange diameter is very important and is given by the formula:

$$m = \frac{d}{D} \tag{7.18}$$

The values of the ratio m are given in Table 7.4

Table 7.4 Values of the ratio of the hole diameter to the flange diameter m

Process used in making hole	Relative thickness of material $(T/d) \times 100\%$								
	2	3	5	8	10	15	20	30	70
Drilling	0.75	0.57	0.48	0.41	0.40	0.34	0.32	0.26	0.22
Punching	0.70	0.60	0.52	0.50	0.50	0.48	0.46	0.45	–

7.5 SPINNING

Spinning is the process of forming a metal part from a circular blank of sheet metal or from a length of tubing over a mandrel with tools or rollers. Fig. 7.13 illustrates representative shapes that can be produced by spinning. The advantages of spinning, as compared to other drawing processes, are the speed and economy of producing prototype samples or small lots, normally less than 1,000 pieces. Tooling costs are less and the investment in equipment is relatively small. However, spinning requires more skilled labor. There are three types of spinning processes: conventional spinning, shear spinning, and tube spinning.

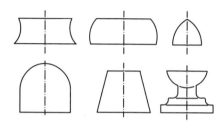

Fig. 7.13 Characteristic shapes of shells formed by spinning

a) Conventional Spinning

In conventional spinning, also called *manual spinning*, a circular blank of sheet metal is held against a form block and rotated while a rigid tool is used to deform and shape the workpiece over the form block. Fig. 7.14 schematically illustrates the manual spinning process. The tools may be operated manually or by a hydraulic mechanism.

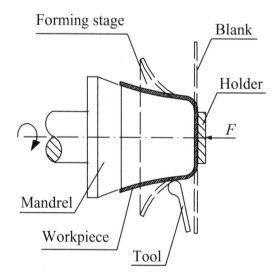

Fig. 7.14 Manual spinning process.

b) Shear Spinning

Shear spinning is the process of forming complex shapes, such as cones with tapering walls; symmetrical-axis curvilinear shapes, such as nose cones; and hemispherical and elliptical tank closures with either uniform or tapering walls. The shape is generated by keeping the diameter of the workpiece constant, as illustrated in Fig 7.15. Although a single roller can be used, two rollers are desirable to balance the radial forces acting on the form block. During spinning, normal wall thickness is reduced. In shear spinning over a conical form block, the thickness T' of the spun part is given by the formula:

$$T' = T \sin \alpha \tag{7.19}$$

where

α = half angle of cone

T = blank thickness

An important factor in shear spinning is the spinnability of the metal. The spinnability is the smallest thickness to which a workpiece can be spun without fracture.

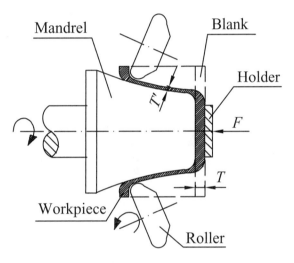

Fig. 7.15 Shear spinning method.

c) Tube Spinning

Tube spinning consists of reducing the thickness of a cylindrical work piece while it is spinning on a cylindrical form block, using rollers (Fig. 7.16). There are two methods: forward and backward. In either example, the reduction in wall thickness results in a longer tube. The ideal tangential force in forward tube spinning may be calculated by the following formula:

$$F_t = Y_{sr} \cdot \Delta T \cdot f \tag{7.20}$$

where:

$$\Delta T = T - T'$$

Y_{sr} = average flow stress of the material

f = feed

Because of friction and other influencing factors, the force exerted is about twice that of the ideal force.

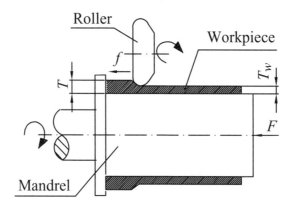

Fig. 7.16 Tube spinning method.

Figure 7.17 shows a relative cost comparison for manufacturing a round sheet metal shell by deep drawing and by manual spinning.

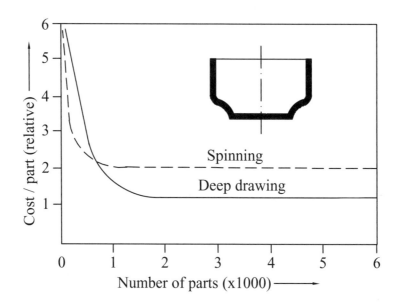

Fig. 7.17 Cost comparison for manufacturing a round sheet shell by conventional spinning and by deep drawing.

7.6 FLEXIBLE DIE FORMING

In the flexible die forming process, one of the dies in a set (punch or die) is replaced with a flexible material such as rubber or polyurethane. Polyurethane is widely used due to its resistance to abrasion and its long fatigue life.

The flexible pad is attached to the ram of the press and can be used to form parts of various shapes. Thus, the production of a specific product only requires one rigid form block. This gives the process a cheapness and versatility ideal for small product series, particularly in such industries as the aircraft industry; about half of all sheet metal parts are made using these processes. The large variety of parts (of different sizes, shapes, and thicknesses), and the relatively small product series of each part, are two important factors of sheet metal parts production in the aircraft industry that make it necessary to use very versatile and low-cost production processes. Several processes utilize flexible pad forming techniques that offer these characteristics, including Guerin, Verson–Wheelen, Marform, and hydroforming processes.

7.6.1 Guerin Process

A typical setup is illustrated in Fig. 7.18, in which a rigid forming block is placed on the lower bed of the press. On top of this block, the blank is positioned and a soft die of rubber or polyurethane (hardness 50 to 70 Shore) is forced over the rigid block and blank into its required shape. The thickness of the rubber pad, which is held in a sturdy cast-iron or steel container, is usually three times the height of the formed block, but it must be at least 1.5 times thicker than the height of a rigid form block. During the process cycle, the rubber pad deforms elastically over the form block and the blank, applying a large pressure, typically of 100 MPa (14,500 psi). The pressure that the soft die exerts on the blank is uniform, so that the forming process creates no thinning of the material, but the radii are shallower than those produced in conventional dies.

There are advantages to the Guerin process besides the relatively low tooling costs and versatility. Due to the softness of the rubber, the top surface of the workpiece will not be damaged or scratched under normal circumstances, many shapes and kinds of workpieces can be formed in one press cycle, and different tooling materials such as epoxy, wood, aluminum alloys, or steel may be used. The process also has some disadvantages, such as a long cycle time and the tendency for the rubber to be easily torn.

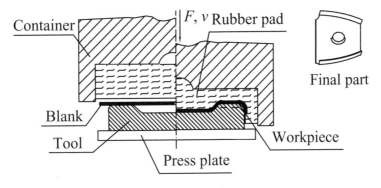

Fig. 7.18 Guerin forming process.

7.6.2 Verson–Wheelon Process

The Verson–Wheelon process is based on the Guerin process. This process uses a fluid cell to shape the blank in place of a thick rubber pad, as illustrated in Fig. 7.19. This method allows a higher forming pressure to be used in forming the workpiece.

The forming principle is simple: Tooling form blocks are placed in a loading tray and sheet metal blanks are placed over the blocks. A throw, usually a rubber pad, is next placed over the blanks to cushion the sharp edges. There, a flexible fluid cell is inflated with high pressure hydraulic fluid. The fluid cell expands and flows downward over and around the die block, exerting an even, positive pressure at all contact points. As a result, the metal blank is formed to the exact shape of the die block. The press is then depressurized for loading the next tray.

This process allows parts to be formed from aluminum, titanium, stainless steel, and other airspace alloys in low volumes. However, parts produced by this process are limited in depth.

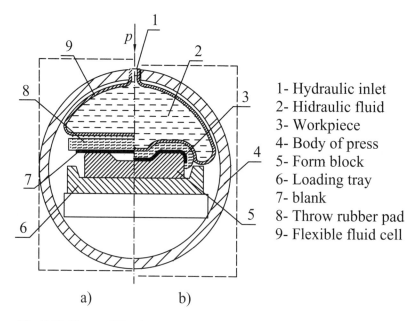

1- Hydraulic inlet
2- Hidraulic fluid
3- Workpiece
4- Body of press
5- Form block
6- Loading tray
7- blank
8- Throw rubber pad
9- Flexible fluid cell

a)   b)

Fig.7.19 Verson–Wheelon process: a) start of process; b) end of forming.

7.6.3 Marform Process

The Marform process is similar to the Guerin process, but is used for deep drawing and forming of wrinkle-free shrink flanges. A sheet metal blank is held firmly between a blankholder and the rubber pad, as shown in Fig. 7.20.

The rubber pad is forced down against a flat blank that rests on a blankholder. A hydraulic servo valve controls counterpressure on the blankholder during the process. As downward pressure mounts, the blankholder is slowly lowered, forcing the sheet metal blank to take the shape of the form block.

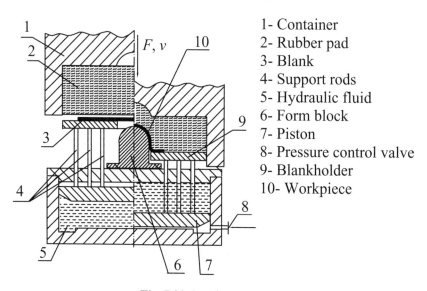

1- Container
2- Rubber pad
3- Blank
4- Support rods
5- Hydraulic fluid
6- Form block
7- Piston
8- Pressure control valve
9- Blankholder
10- Workpiece

Fig. 7.20 Marform process.

7.6.4 Hydroforming Process

The hydroforming process has been well known for the last 20 years and has undergone extremely swift development in aircraft and automotive applications, especially in Germany and the United States.

Hydroforming is a manufacturing process in which fluid pressure is applied to a ductile metallic blank to form a desired workpiece shape. Hydroforming is broadly classified into *sheet* and *tube hydroforming*. Sheet hydroforming is further classified into sheet hydroforming with a punch and sheet hydroforming with a die, depending on whether a male (punch) or a female (die) tool will be used to form the workpiece.

The principle of the punch sheet metal hydroforming process is illustrated in Fig. 7.21a. The blankholder is provided with a seal; a container maintains the pressure medium, which is usually a water-oil emulsion. A hydraulic servo valve controls the counterpressure during the process. After the blank is placed on the die, the blankholder presses the sheet blank, and the punch forms the blank against the medium, creating pressure that is controlled through the punch stroke.

The principle of sheet metal hydroforming using a die is illustrated in Fig. 7.21b. The process has two phases:

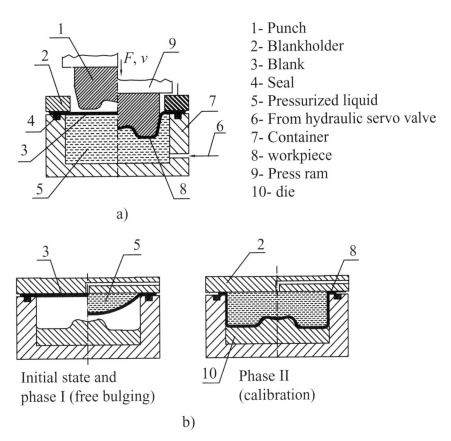

1- Punch
2- Blankholder
3- Blank
4- Seal
5- Pressurized liquid
6- From hydraulic servo valve
7- Container
8- workpiece
9- Press ram
10- die

a)

Initial state and
phase I (free bulging)

Phase II
(calibration)

b)

Fig. 7.21 The principle of sheet metal hydroforming: a) with a punch, b) with a die.

Phase I. Under hydropressure, the sheet metal is deformed freely into the die cavity until it contacts the die surface. Free forming ensures uniform deformation in the sheet metal, which reduces the tendency for blank tearing, caused by localized deformation, and improves dent resistance of the hydroformed part as compared to a conventionally formed part. Phase one is finished after a large portion of the workpiece has made contact with the die surface. At that moment of contact, friction is increased and the flow of metal is reduced.

Phase II. In this phase, the workpiece is calibrated against the die cavity to obtain the final desired shape. At this time, it is necessary to use a high fluid pressure.

The working medium pressure in the hydroforming process varies from 5 to 100 MPa (725 to 14,500 psi); during the forming process, this pressure is adjusted as follows according to the geometry and material of the workpiece:

- aluminum: 5 to 20 MPa
- steel: 20 to 60 MPa
- stainless steel: 30 to 100 MPa

Sheet metal hydroforming processes are mostly used for deep drawing. However, the process can also be used for stretch forming.

Compared to conventional deep drawing, hydroforming deep drawing has some advantages:

- The deep drawing ratio can be increased.
- The tool costs are reduced.
- Material thinout is minimal and is usually less than 10 percent.
- Greater versatility in forming complex shapes and contours is possible.
- The outer surface of the workpiece is free from defects.
- Low work hardening occurs.
- Fewer operations are required.

Although various types of flexible die forming operations have been used for years in the aircraft industry, sheet metal hydroforming is particularly well-suited for the prototyping and low-volume production required by the industry.

7.6.5 Force in Die Forming

The press force for flexible die forming is given by the following formula:

$$F = p \cdot A \tag{7.21}$$

where

A = area of the pad of rubber

p = specific pressure (Table 7.5)

A specific pressure (MPa) for drawing a workpiece of aluminum alloy is given in Table 7.5.

Table 7.5 Values of specific pressure p (MPa)

Ratio $m= d/D$	Relative thickness of material $Tsr = (T/D) \times 100\%$			
	1.3	1.0	0.66	0.44
0.6	25.5	27.5	32.0	35.5
0.5	27.5	29.5	34.0	37.0
0.44	29.5	32.0	34.4	39.3

Table 7.6 provides given maximum values for the drawing ratio m, the drawing height in one operation h, and the minimum drawing radius R_m, for different kinds of materials.

Table 7.6 Maximum values for drawing ratio m, drawing height h, and minimum drawing radius Rm

MATERIAL	M	h	Rm
Aluminum	0.45	$1.0d$	$1.5T$
Al-alloy	0.50	$0.75d$	$2.5T$
Low-carbon steel	0.50	$0.75d$	$4.0T$
Stainless steel	0.65	$0.33d$	$2.0T$

The maximum drawing height h of rectangular and square-drawn shells in one drawing operation is given by the following:

$$h \le R_e - \text{for steel}$$
$$h \le 3.5R_e - \text{for aluminum}$$

where

R_e = drawing corner radius

Part
Three

Die Design

This part of the text is intended to present as complete a picture as space will permit of the knowledge and skills needed for the effective design of dies for sheet metal forming. The discussion is presented in six chapters, with detailed discussions of die design for each of the manufacturing processes covered in Part II. Chapter 13 gives data on basic tool and die materials, their properties and applications. Special pay attention is given to the basic functions of work and die components, their design and necessary calculations. Although many examples are included, it should be evident that it is not possible to present all the data, tables, and other information needed to design complicated tools, dies, etc. in one text. The tool and die designer must have tables, handbooks, and literature available to be completely effective as a die designer. This part presents most of the practical information needed by a designer.

Eight

Basic Die Classifications and Components

8.1 DIE CLASSIFICATIONS

Dies can be classified according to a variety of elements and in keeping with the diversity of die designs. This chapter discusses primarily die classifications depending on the production quantities of stamping pieces, (whether high, medium, or low), the number of stations, and according to the manufacturing processes.

8.1.1 Die Classifications Depending on the Production Quantity of Parts

Depending on the production qualities of pieces—high, medium, or low—stamping dies can be classified as follows:

Class A. These dies are used for high production only. The best of materials are used, and all easily worn items or delicate sections are carefully designed for easy replacement. A combination of long die life, constant accuracy throughout the die life, and ease in maintenance are prime considerations, regardless of tool cost.

Class B. These dies are applicable to medium production quantities and are designed to produce the designated quantity only. Die cost as related to total production becomes an important consideration. Cheaper materials may be used, provided they are capable of producing the full quantity, and less consideration is given the problem of ease of maintenance.

Class C. These dies represent the cheapest usable tools that can be built and are suitable for low-volume production of parts.

8.1.2 Die Classifications According to Number of Stations

According to the number of stations, stamping dies may be classified as:

- Single station dies
- Multiple station dies

a) Single Station Dies

Single station dies may be either compound dies or combination dies.

Compound die. A die in which two or more cutting operations are coordinated to produce a part at every press stroke is called a compound die.

Combination die. A die in which both cutting and noncutting operations are coordinated to produce a part at one stroke of the press is called a combination die.

b) Multiple Station Dies

Multiple station dies are arranged so that a series of sequential operations is coordinated with each press stroke. Two die types are used:

- Progressive dies
- Transfer dies

Progressive die. A progressive die is used to transform coil stock or strips into a completed part. This transformation is performed incrementally, or progressively, by a series of stations that cut, form, and coin the material into the desired shape. The components that perform operations on the material are unique for every part. These components are located and guided in precision cut openings in plates, which are in turn located and guided by pins.

The entire die is actuated by a mechanical press that moves the die up and down. The press is also responsible for feeding the material through the die, progressing it from one station to the next with each stroke.

Transfer die. In transfer die operations, individual stock blanks are mechanically moved from die station to die station within a single die set. Large workpieces are done with tandem press lines where the stock is moved from press to press in which specific operations are performed.

8.1.3 Die Classifications According to Manufacturing Processes

According to the manufacturing processes, stamping dies may be classified as:

    a) Cutting dies

    b) Bending and forming dies

    c) Drawing dies

    d) Miscellaneous dies

a) Cutting Dies

Cutting dies may be either blanking dies or punching dies.

Blanking dies. A blanking die produces a blank or workpiece by cutting the entire periphery in one simultaneous operation. By construction, they can be as cutoff die, drop-through die, and return blanking die.

Punching dies. Punching die punches holes in a workpiece as a separate operation. There are two principal reasons for punching holes in a separate operation instead of in combination with other operations: when a subsequent bending, forming, or drawing operation would distort the previously punched hole or holes; or when the edge of the punched hole is too close to the edge of the die section.

b) Bending and Forming Dies

Bending die deforms a portion of a flat blank to some angular position. The line of bend is straight along its entire length. The operation of forming is similar to that of bending, except that the line of deformed is curved instead of strength, and plastic deformation in the workpiece material is more severe.

c) Drawing Dies

A drawing die draws the blank into a cup or another differently shaped shell. Drawing dies can be classified by their design as: drawing dies for single-action press and drawing dies for double-action press.

Drawing dies for single-action press. The blank is placed on the pressure of the drawing die. Descent of the upper die causes the blank to be gripped between the surface of the pressure pad and the lower surface of the die ring. Further descent of the ram causes the blank to be drawn over the punch until it has assumed the cup shape.

Drawing dies for double-action press. The die consist of the upper shoe and the lower shoe. A blankholder is attached to the upper shoe and the workpiece ejector is located into the lower shoe. The punch is attached directly to the inner slide of the press, and the upper shoe with a blankholder to the outer press slide. Workpiece ejector is powered by a mechanism located below the bed of the press.

d) Miscellaneous Dies

Miscellaneous dies can be classified by their spatial operations or design as: trimming dies, shaving dies, curling dies, bulging dies, necking dies, extruding dies, assembly dies, and side-cam dies.

Trimming dies. A trimming die cuts away portions of a formed or draw shell that have become irregular. The edge quality requirement of the piece and size will indicate which type of trimming die must be designed.

Shaving dies. A shaving die removes a small amount of metal from around the edges of a blank or hole in order to improve the surface and accuracy.

Curling dies. A curling die forms the sheet metal at the edge of a workpiece into a circular shape or hollow ring. Most often, curling is applied to the open edges of cylindrical cups and other drawn but not cylindrical shells.

Bulging dies. A bulging die expands a portion of a drawn shell or tube, causing it to bulge. The upper part of the punch is a rubber ring within which is positioned a spreader rod.

Necking dies. The operation necking is exactly the opposite of bulging. A necking die reduces a portion of the diameter of the deep drawn shell or tube, making the portion longer as well.

Expanding dies. An expanding die is commonly used to enlarge the open end of a drawn shell or tubular stock with a punch.

Assembly dies. An assembly die assemble two or more parts together by press-fitting, riveting, staking, or other means.

Side-cam dies. A side-cam die transforms vertical motion from the press ram into horizontal or angular motion; it can perform many ingenious and complicated operations.

8.2 BASIC DIE COMPONENTS

The components that are generally incorporated in a punching or blanking die with a guide plate are shown in Fig. 8.1, and in Fig 8.2 with a guide system (guideposts and guidepost bushings). These figures show the die in the conventional closed position.

   The upper die subset in Fig 8.1 is made up of the following components: the punch holder, the backing plate, the punch plate, the punch, and the shank. The components are fastened together with screws and a dowel at the die subset, which is attached to the ram of the press.

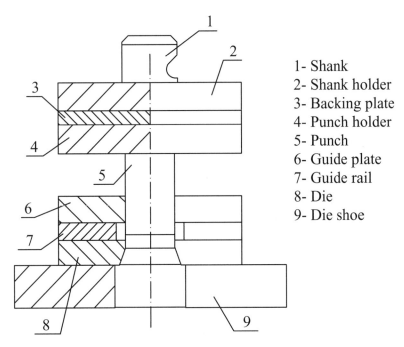

Fig. 8.1 Die with guide plate.

1- Shank
2- Shank holder
3- Backing plate
4- Punch holder
5- Punch
6- Guide plate
7- Guide rail
8- Die
9- Die shoe

The lower die subset is made of the following components: the die shoe, the die (also called die block), the guide rails, and the guide plate. The components also are fastened with screws and dowels at the die subset, which attaches to the bolster plate.

Figure 8.2 shows a die set with the basic die components and the guide system, which consists of the guide posts and guide post bushings.

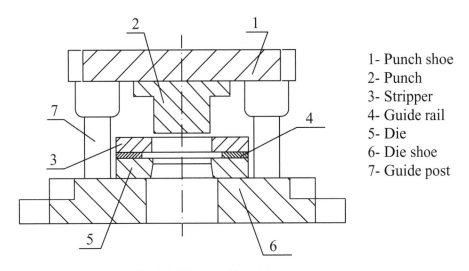

1- Punch shoe
2- Punch
3- Stripper
4- Guide rail
5- Die
6- Die shoe
7- Guide post

Fig. 8.2 Die set with guide post system.

According to the function of the die, all components may be classified into two groups:

Technical components. The technological components directly participate in forming the workpiece, and they have direct contact with a material. Examples include the punches, die block, guide rails, form block, drawing die, stripper, and blank holder.

Structural components. The structural components securely fasten all components to the subset and die set. They include the die set, the punch holder, the die shoe, the shank, the guideposts, the guidepost bushings, the springs, screws, and dowels.

Die sets are used to hold and maintain alignment of the technological components. The die set consists of the upper shoe, lower shoe, and guide system (guide posts and guide bushings). Figure 8.3a shows various guide post positions depending upon the type of operation to be done. These designs are referred to as two- or four-guide post systems. A die set and its components are shown in Fig. 8.3b, in which the elevations have been drawn according to the conventions used by die designers. It consists of the punch shoe (usual upper shoe), die shoe, guide post bushings, and guide posts.

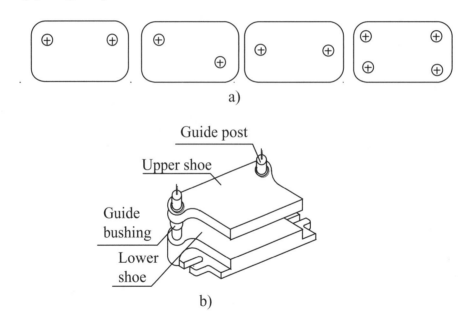

Fig. 8 3 Die set: a) various post positions; b) die set and its components.

Figure 8.4 shows one type of guide post and guide post bushing. The guide post has two different diameters — the larger diameter is for fastening to the die shoe whereas the smaller diameter is for sliding at the guide post bushing. The joint between the die shoe and the guide post is a press fit (usually H7/p6 or H6/p5); the joint between the guide post bushing hole and the guide post is a sliding fit (H7/h6 or G7/h6).

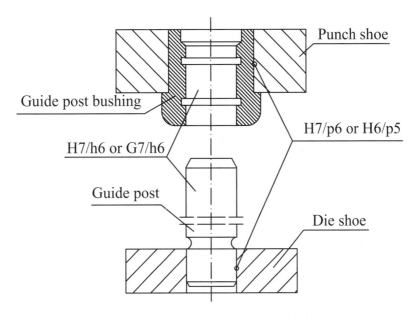

Fig. 8.4 Guide post and guide post bushing.

Other mechanical methods may be used to join the die shoe and guide post. Two types (press fit and slide fit) are shown in Fig. 8.5.

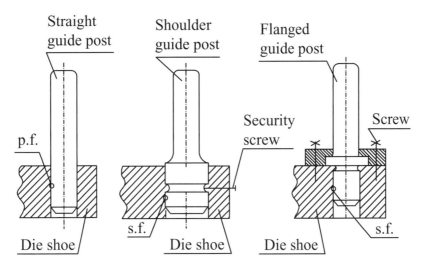

Fig. 8.5 Two examples for joining the die and guide post.

The guide post holes may be made directly in the punch shoe or as separate guide post bushings that are press fitted in the punch shoe (Fig. 8.3). Inside the guide post bushing is a channel for lubrication (sometimes the channel is made in the guide post). The best guidance and the most precision are achieved if a ball cage die set is used, as shown in Fig. 8.6.

To make sure that the setup is correct and to avoid wrong positioning, the guide posts and guide post bushings on one side must be of a different diameter from the guide posts and bushings on the other side.

The preceding information is generally valid for die components for all sheet-metal forming processes; in the following chapters, more specific information concerning die components for each individual type of sheet-metal forming process is presented.

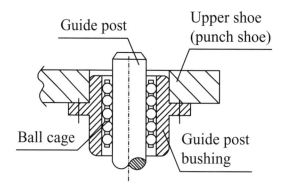

Fig. 8.6 Ball cage die set.

Nine

Blanking and Punching Dies

9.1 INTRODUCTION

Blanking and punching dies are known as cutting dies. They may be simple, combination, or compound. *A blanking die* is generally cheaper to make and faster in operation than a trim die. A single blanking die can produce either a right or left part, whereas two trim dies are needed for trimming: one die for right-hand parts and another die for left-hand parts. When a sheared flat blank drops through the die block (die shoe), it piles up on top of the bolster plate. If the blank goes through the hole, it is called a *drop-blank die*. A die in which the sheared blank returns upward is called a *return-blank die*. Return-blank dies are slower in operation and cost more to build than drop-blank dies.

A punching die is a typical single-station die design for production holes made in flat stock, which may be manually or automatically fed. The stock guide keeps the stock on a straight path through the die. The amount of stock travel is controlled by the method of feeding, by stops of various designs, or by direct or indirect piloting. *A combination die* is a single-station die in which both cutting and non-cutting operations are completed at one press stroke. *A compound die* is a single-station die in which two or more cutting operations are completed at every press stroke.

In this chapter, the design characteristics of cutting dies, as well as the very important calculations for the technological parts of cutting dies, are described.

9.2 DIE BLOCKS

A die block is a technological component that houses the opening and receives punches. These die openings may be machined from a solid block of tool steel or may be made in sections. The die block is pre-drilled, tapped, and reamed, before being fastened to the die shoe.

9.2.1 Die Opening Profile

Die opening profiles depend on the purpose and required tolerance of the workpiece. Four opening profiles are shown in Fig. 9.1. The profile in Fig. 9.1a gives the highest quality workpiece, but its machining is the most expensive. Die blocks must frequently be resharpened to maintain their edge. To allow a die block to be sharpened more times, the height of the die block h needs to be greater than the thickness of the workpiece. The value of h is given in Table 9.1. This kind of die opening is used for blanking parts having complex contours with greater accuracy.

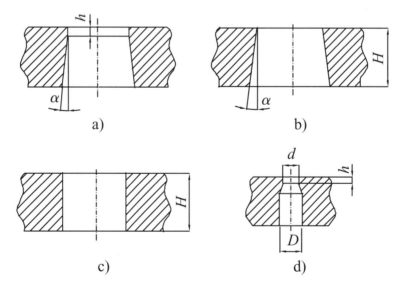

Fig. 9.1 Types of die opening profile.

Table 9.1 Value of Dimension *h* Depends on Material Thickness *T*

Material thickness *T* (mm)	< 0.5	0.5 to 5.0	5.1 to 10.0
Height *h* (mm)	3.0 to 5.0	5.1 to 10.0	10.1 to 15.0
Angle α	3^0 to 5^0		

The die opening profile in Fig. 9.1b is used for making small parts with low accuracy. The angle of the cone, $\alpha = 10'$ to $20'$ for material of thickness $T < 1$ mm, and $\alpha = 25'$ to $45'$ for material of thickness $T \geq (1$ to $5)$ mm. For the angle to be correctly derived, the following relationship must be satisfied:

$$\alpha \leq \text{arctg} \frac{\Delta}{2H}$$

where

Δ = tolerance of workpiece

H = height of die block

The simplest die opening profile is the cylinder, as shown in Fig. 9.1c. This type of profile is used for making relatively large parts. With this profile, after the part is cut out, it is pushed up and away from the die.

The two-cylinder die opening profile, shown in Fig. 9.1d, is used to punch small-diameter ($d \leq 5$ mm) holes. The value of *h* can be taken from Table 9.1. The diameter of the larger cylinder needs to be $D = d + 3$ mm.

9.2.2 Fastening to the Die Shoe

There are many methods for fastening a die block to a shoe. Figure 9.2a shows one method, in which socket head screws are inserted from the bottom of the die shoe into threaded holes in the die block. Dowels are used to prevent a shift in the position of the block.

Sometimes the die opening is made from a bushing and inserted into a machine steel retainer. If the bushing has a shoulder, it is held in the retainer, as shown in Fig. 9.2b. If it has no shoulder, it is pressed into the retainer, as shown in Fig. 9.2c. The lower end of the bushing has a reduced diameter to insure alignment when it is pressed into the retainer. The bushing can be fastened into the retainer with a ball and screw, as shown in Fig.9.2d. The purpose is to save the cost of large amounts of tool steel. This type of fastening is generally used to allow replacement of the die ring if it is worn or damaged. Sometimes a bushing is used at the bottom, as is shown in Fig. 9.2e. The purpose is to save the cost of large amounts of tool steel.

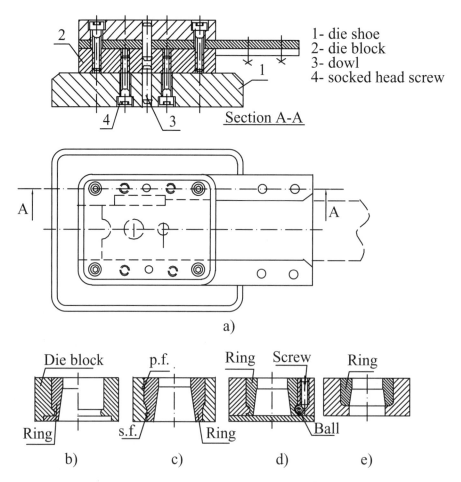

1- die shoe
2- die block
3- dowl
4- socked head screw

Section A-A

a)

Die block   p.f.   Ring   Screw   Ring

Ring   s.f.   Ring   Ball

b)   c)   d)   e)

Fig. 9.2 Fastening to the die shoe: a) fastening a die block to a shoe; b), c), d) and e) methods of inserting a bushing into a machined retainer.

9.2.3 Sectioned Die

If a workpiece is large, or if the die opening is complicated, and the contours are difficult to machine, the die may be made in sections. If possible, the die sections should be of approximately the same dimensions to ensure economic use of the tool steel. Several types of sectional dies and punches are shown in Fig. 9.3. It is also possible to use a bevel-cut angle on the face of a sectional die. The bevel shear may be convex or concave (see Fig. 4.6).

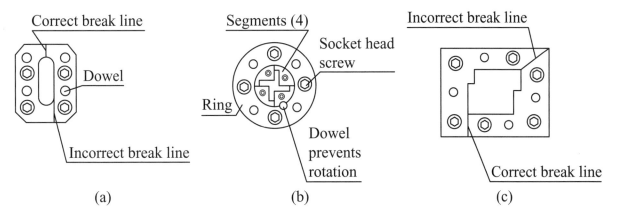

Fig. 9.3 Types of sectioned dies: a) and c) correct and incorrect ways of applying break lines in sectioned die; b) round sectional die with employing ring to retain four segments.

To save very expensive tool steel, a die and punch with welded edges is often used (as shown in Fig 9.4), primarily for blanking parts of larger dimensions and material thickness up to $T = 1.5$ mm. The die is made of carbon alloy steel, and the edges are welded with a special electrode of alloy steel, and then machined. The welded edge may have a hardness of up to 60 Rc. This type of die is cheap, and repair is easy.

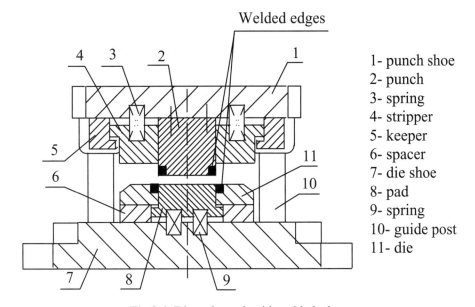

1- punch shoe
2- punch
3- spring
4- stripper
5- keeper
6- spacer
7- die shoe
8- pad
9- spring
10- guide post
11- die

Fig.9.4 Die and punch with welded edges.

Carbide dies are widely used to produce small electrical parts at lower cost per piece compared with steel dies. When a carbide insert is subjected to high-impact load, it must be supported externally by pressing the carbide ring into a hardened steel holder. Suitable steels for die holders include SAE 4140, 4340, and 6145 hardened to 38 to 42 Rc. A ratio of 2:1 between the case and the carbide insert has been satisfactory for most applications, but the ratio can be less for light work.

9.2.4 Calculation of Die Block Dimensions

A die block for blanking and punching operations is loaded at force *F.* About 40 percent of this force is exerted in a way that would fracture the die block in the radial plane. However, the die block is additionally loaded with the friction force produced when the blanked or punched material is pushed through the opening of the die. The calculations for the die block dimensions are very often simplified, making use of two empirical formulas that calculate only the thickness of the die block *H* and the width of the wall *e* (Fig. 9.5). The height (thickness) *H* of the die block is calculated by the formula:

$$H = \left(10 + 5T + 0.7\sqrt{a+c}\right)c \qquad (9.1)$$

where

    T = material thickness

    a, b = opening die dimensions

    c = constant whose value depends on the mechanical properties of the workpiece material and is given in Table 9.2

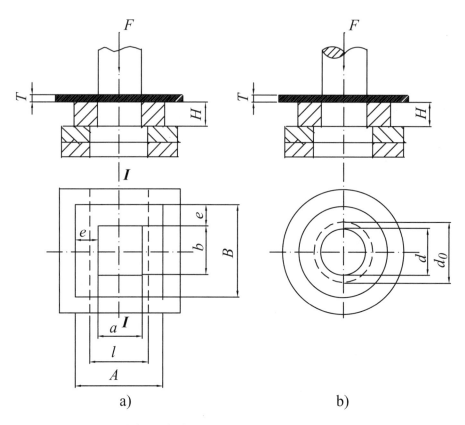

a)   b)

Fig. 9.5 Schematic for calculation of die block dimensions.

Table 9.2 Value of Constant *c*

UTS (MPa)	117	245	392	784
c	0.6	0.8	1.0	1.3

The wall thickness *e* is given by the formula:

$$e = (10 \text{ to } 12) + 0.8H \tag{9.2}$$

If the opening die has a contour with an angle less than 90 degrees, the value of *e* needs to be increased from 15 to 20 percent.

The dimensions of the rectangular die block in Fig. 9.5a are:

$$A = a + 2e$$

$$B = b + 2e$$

The maximal bend moment for a rectangular die block is:

$$M = \frac{1}{8} Fl \tag{9.3}$$

The resistance moment at section I-I is:

$$W = \frac{(B-b)H^2}{6} \tag{9.4}$$

The stress bending of the die block is:

$$\sigma_s = \frac{M}{W} = \frac{6Fl}{8(B-b)H^2} \tag{9.5}$$

For a rectangular die block, the following formula must be applied:

$$\sigma_s = 0.75 \frac{Fl}{(B-b)H^2} \leq \sigma_{sd} \tag{9.5a}$$

where

σ_{sd} = permitted bending stress

For heat-hardened alloy tool steels, the value of

σ_{sd} = 490 MPa

If the die block has a shape similar to that in Fig. 9.5b, the bending stress may be calculated by the formula:

$$\sigma_s = \frac{2.5F}{H}\left(1 - \frac{2d}{3d_0}\right) \leq \sigma_{sd} \tag{9.5b}$$

This method of calculation is approximate, but for practical use, it is good enough. More precise calculations are based on mechanical theory.

Example:

For example, you may be required to calculate the dimensions of the rectangular die block of heat treated alloy tool steel. The die block is supported by two supports at a distance of $l = 120$ mm. The workpiece material has a thickness of $T = 3$ mm and $UTS = 372$ MPa. The workpiece dimensions are $a = 100$ mm and $b = 120$ mm.

Solution:

The height of the die block is:

$$H = \left(10 + 5t + 0.7\sqrt{a+b}\right) \cdot c$$

$$H = \left(10 + 5(3) + 0.7\sqrt{100+200}\right)(1.0) = 37.2 \text{ mm}$$

The height of the die block can be rounded to $H = 40$ mm.
The wall thickness e is:

$$e = \left(10 \text{ to } 12\right) + 08H$$

$$e = 12 + 0.8(40) = 12 + 32 = 44 \text{ mm}$$

The value of e may be rounded to $e = 50$ mm.
The die block dimensions are:

$$A = a + 2e = 100 + 2(50) = 200 \text{ mm}$$

$$B = b + 2(e) = 200 + 2(50) = 300 \text{ mm}$$

The blanking force is:

$$F = 2(a+b)T0.8(UTS)$$

$$F = 2(100+200)(3)0.8(372) = 53.662 \text{ kN}$$

The bending stress is:

$$\sigma_s = 0.75\frac{Fl}{(B-b)H^2}$$

$$\sigma_s = 0.75\frac{53.662(120)}{(300-200)40^2} = 402.6 \text{ MPa} < \sigma_{sd}$$

These dimensions are acceptable because σ_s 402.6 MPa $< \sigma_{sd} = 490.5$ MPa.

9.3 PUNCHES

Standard punches are available for a wide variety of round, oblong, and square holes. The manufacturers furnish these punches in standard sizes as well as to special order. The main considerations when designing punches are 1) they should be designed so that they do not buckle, 2) they should be strong enough to withstand the stripping force, and 3) they should not be able to rotate as a result of the cutting action.

9.3.1 Punch Face Geometry

It is possible to control the area being sheared at any moment by making the punch and die surface at an angle (beveled), as shown in Fig. 4.6. In Fig. 9.6, several types of punch face geometry are shown:

- flat face surface
- concave face surface
- bevel face surface
- double bevel face surface

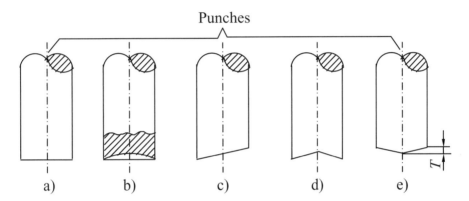

Fig. 9.6 Types of punch face geometry: a) flat, b) concave, c) bevel, d) and e) double bevel.

If the surface of the punch and die are flat, the punch force builds up rapidly during cutting because the entire cross-sectional area is being cut at one time. The punch face geometry in Fig. 9.6 is particularly suitable, with an adequate shear angle on the die, for shearing thick blanks because it reduces the force required at the beginning of the stroke. The angle also reduces the noise level. Typical combinations are:

- flat punch — double bevel die
- concave punch — flat die
- bevel punch — flat die
- flat punch — concave die

It is possible to use other combinations depending on the purpose of the die. When blanking soft and thin material, a tubular punch without a die block may be used.

9.3.2 Methods for Assembling Punches

There are many methods for assembling blanking and piercing punches on a punch holder. Figure 9.7 shows several forms at one end of the punches for such assembly.

Piercing punches are smaller in cross-section and generally longer than blanking punches. They must be designed to withstand shock and buckling. Because of the high probability of damage, they must be designed so that they can be easily removed and replaced. Deflection or buckling of punches may be avoided by making the body diameter of the punch larger than the cutting diameter or by guiding the punch through a bushing, as shown in Fig. 9.7d and Fig. 9.7e.

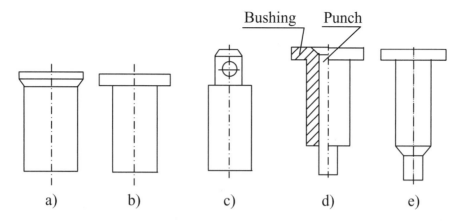

Fig. 9.7 Forms at one end of piercing punches.

Sometimes it is necessary to insert a hardened backing plate between the head of the punch and the punch holder. Whether or not it is necessary to use a backing plate is dependent on the specific pressure between the head of the punch and the punch holder. If the following condition is satisfied,

$$p = \frac{F}{A} = \frac{4F}{\pi d^2} < p_{\mathrm{d}} = 245 \text{ MPa}$$

a backing plate is not necessary (Fig. 9.8a). However, a backing plate is necessary under the conditions illustrated in Fig. 9.8b.

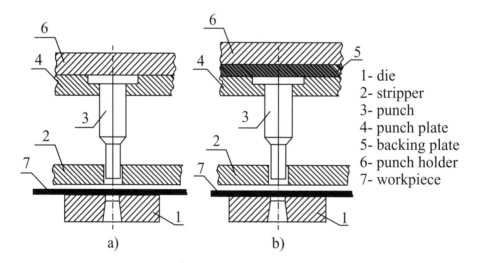

1- die
2- stripper
3- punch
4- punch plate
5- backing plate
6- punch holder
7- workpiece

Fig. 9.8 Assembling punch: a) without backing plate, b) with backing plate.

Because of the high probability of damage to piercing punches, they must be designed so that they can be easily removed and replaced. Figure 9.9 shows four methods of designing such punches, which all allow for quick removal and replacement at the punch plate.

Fastening with ball and screw. This type of fastener is used with complex dies (Fig. 9.9a).

Fastening with a ball under spring pressure. The punch can be released by pushing the ball through hole *a*. This type of fastener (Fig. 9.9b) is used on dies for punching holes of *d* = 3 to 30 mm in material thickness *T* = up to 3 mm.

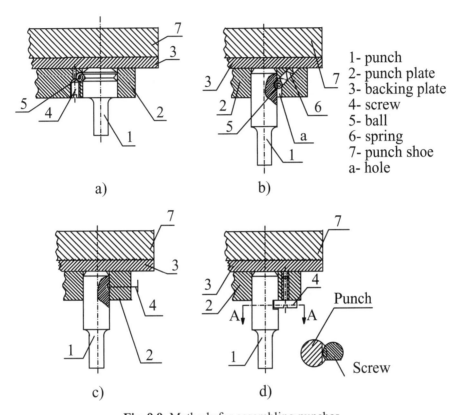

1- punch
2- punch plate
3- backing plate
4- screw
5- ball
6- spring
7- punch shoe
a- hole

Fig. 9.9 Methods for assembling punches.

Fastening with a screw. Screw fasteners (Fig.9.9c) are used for simple dies, and the punch is fastened directly on the ram of the press. The domain of application is for punching holes *d* = 20 to 60 mm.

Fastening with a specially shaped screw head. When the screw is turned 90 degrees (Fig.9.9d), the punch is freed. Sometimes, it is more economical to use a machine steel spacer and a tool steel plate or ring for a punch than it is to make a large blanking punch out of one piece of tool steel.

At other times, it may be easier and more economical to make punches out of sections and fit them into the desired pattern than it is to attempt to make the punch from one piece. Such a design may be desirable when the punches need to be large or irregularly shaped. Another advantage is that a section may be removed and replaced if it becomes worn or broken.

9.3.3 Punch Calculations

By using the ratio between the length and the cross-section area of the punch, the dimensions of the punch may be calculated in two ways: Compression stress and buckling calculation.

a) Compression Stress

Compression stress for short punches may be calculated by the formula:

$$\sigma_d = \frac{F}{A} \leq \sigma_{pd} \tag{9.6}$$

where

F = punching force

A = punch cross section area

σ_{pd} = permissible compression stress

For hardened tool steels, σ_{pd} = 980 to 1560 MPa.

b) Buckling Calculation

If the punch is fixed at one end, as shown in Fig 9.10a, the critical force should initially be calculated using the Euler formula:

$$F_{cr} = \frac{\pi^2 EI_{min}}{4l} \tag{9.7}$$

where

I_{min} = minimal second moment of area

l = free length of punch

E = modulus of elasticity

If critical force F_{cr} equals punch force F, then the maximum length of the punch may be calculated by the following formula:

$$l_{max} = \sqrt{\frac{\pi^2 EI_{min}}{4LT(0.8UTS)}} \tag{9.8}$$

where

L = length of cut

T = thickness of material

For a punch fixed at one end and guided at the other end, as shown in Fig. 9.10b, the critical force may be calculated by the formula:

$$F_{cr} = \frac{2\pi^2 EI_{min}}{l^2} \tag{9.9}$$

The critical force exerted by a guided punch is 8 times greater than that exerted by a free-end punch. Consequently, the maximal length of a guided punch is $\sqrt{8} = 2.8$ times larger than that of a free end punch. Special attention needs to be paid to the design of the die for punching small-diameter holes in thick material because greater than allowable compression stress in the punch may occur. For punching any material where the shearing stress is $\tau \geq 295$ MPa, the punch diameter must be greater than the thickness of the workpiece.

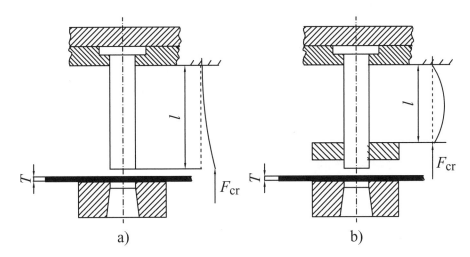

Fig. 9.10 Different end conditions: a) one end is fixed, and the other end is free, b) one end is fixed, and the other end is guided.

9.4 STRIPPER PLATES

When a punch shears its way through a work piece, the material contracts around the punch to the degree that it takes a substantial force to withdraw the punch from the material.

Efficient removal of the workpiece and scrap from the die increases productivity, quality, and workplace safety.

9.4.1 Stripper Force

The force required to strip the material from the punch, as shown in Fig. 9.11, may be calculated by the following equation:

$$F_s = C_s F \tag{9.10}$$

where

C_s = stripping constant (Table 9.3)
F = punch force

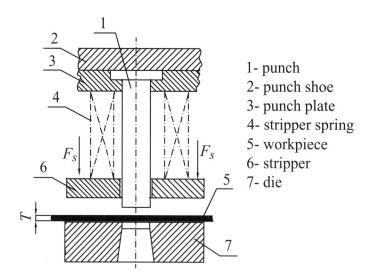

1- punch
2- punch shoe
3- punch plate
4- stripper spring
5- workpiece
6- stripper
7- die

Fig. 9.11 Schematic illustration showing positioning of the stripper.

Table 9.3 Values of C_s

MATERIAL THICKNESS T (mm)	Types of work processes		
	Simple punching or blanking	Compound punching or blanking	Punching and blanking at same time
Up to 1.0	0.02 to 0.06	0.06 to 0.08	0.10 to 0.12
1.0 to 5.0	0.06 to 0.08	0.10 to 0.12	0.12 to 0.15
Over 5.0	0.08 to 0.10	0.12 to 0.15	0.15 to 0.20

9.4.2 Stripper Design

When a punch operates its way through workpiece material, the material contracts around the punch; it will be carried upward when the punch ascends unless there is some device, called a *stripper*, to prevent this.

Generally, there are two types of strippers: the solid stripper and the elastic stripper.

a) Solid Stripper

The cheapest and simplest design for a stripper is shown in Fig. 9.12. Two additional methods for mounting solid stripper plates are shown in Fig 9.13. In Fig. 9.13a, a spacer is provided to raise the stripper so that there is clearance between the work strip and the stripper plate. In Fig. 9.13b, the stripper plate has had a channel milled into its lower surface.

In all types of solid strippers for stripping scrap or workpieces, the force of the press is used for the stripping operation.

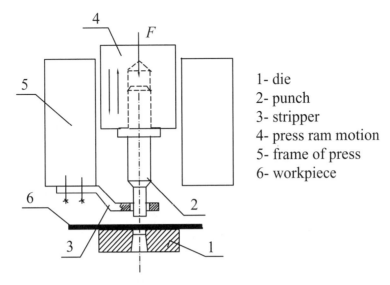

Fig.9.12 Schematic illustration of mounting a simple solid stripper.

1- die
2- punch
3- stripper
4- press ram motion
5- frame of press
6- workpiece

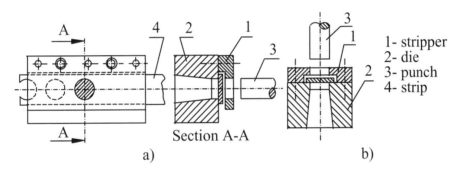

Section A-A

a)   b)

1- stripper
2- die
3- punch
4- strip

Fig. 9.13 Schematic illustration of mounting a solid stripper: a) stripper fixed on one side; b) stripper fixed on two sides.

b) Elastic Stripper

Sometimes it is desirable to hold the scrap strip in a flat position before the punch makes contact with the workpiece. This arrangement is advisable when it is necessary to be very accurate, when punching very thin material, or when thin punches are used. These types of strippers use a compression spring or rubber pad to produce the stripping force. The stripper is generally made from a plate that provides the desired configuration and is suspended from the punch holder with stripper bolts and compression springs. An example of such a stripper plate is shown in Fig. 9.14.

There are many ways of retaining springs. In Fig. 9.14, both the punch holder and the stripper plate are counterbored to provide retainers. However, a counterbore in the stripper plate is often not possible.

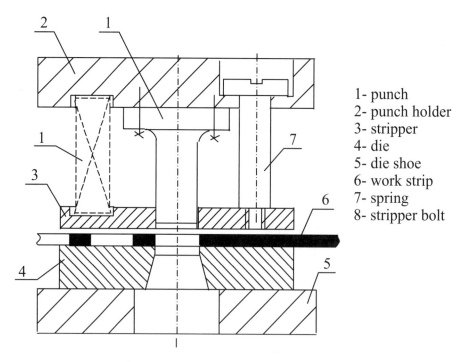

1- punch
2- punch holder
3- stripper
4- die
5- die shoe
6- work strip
7- spring
8- stripper bolt

Fig. 9.14 Blanking die with elastic stripper.

Where it is not desirable that the stripper plate be counterbored, a spring pilot may be used. Also, the counterbore may be in the punch plate rather than in the punch holder. In some instances, small hydraulic cylinders are used instead of springs to produce the force necessary to strip the scrap from the punch. Helical springs, shown in Fig. 9.15, are used to produce the force necessary to strip the scrap strip from the punch.

The maximum static force F_{max} for the helical spring may be calculated with the following formula:

$$F_{max} = \frac{\pi d^3 \tau}{8D} \tag{9.11}$$

where

    d = diameter of wire

    D = outside diameter of helical spring

    τ = shearing stress (τ = 490 to 685 MPa)

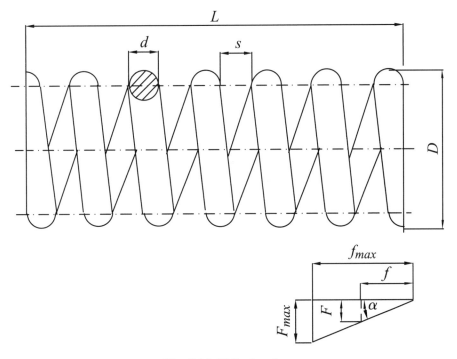

Fig. 9.15 Helical spring.

The total deflection of the spring may be calculated by the formula below:

$$f_{max} = \frac{8nD^3 F_{max}}{Gd^4} \tag{9.12}$$

where

    G = modulus of elasticity in the shear

    n = number of active coils

The average value of G for steels used for springs is $G = 75{,}000$ to $83{,}000$ *MPa*.

    The spring index is found from:

$$c = \frac{F_{max}}{f_{max}} = \frac{F}{f} = \text{tg}\,\alpha \tag{9.13}$$

Therefore, the spring force is:

$$F = cf = F_{max}\frac{f}{f_{max}} \tag{9.14}$$

    The springs are equally spaced around the center of pressure of the punches. During assembly, the springs need to be preloaded with a force of (0.1 to 0.2) F_{max}. The length of the spring may be calculated by means of the formula:

$$L = (n + 1.5)d + ns \tag{9.15}$$

where

  s = distance between two coils (s_{min}= 0.1d at maximum loaded spring)

If rubber pads, as shown in Fig. 9.16, are used to produce the force necessary to remove the scrap strip from the punch, the rubber pads should have a minimum hardness of 68 Shore. The rubber pad has a cylindrical shape with a height-to-diameter ratio of 0.5 to 1.5. Maximum deflection is:

$$f_{max} = (0.35 \text{ to } 0.40)h \qquad (9.16)$$

The stripper force may be calculated by this formula:

$$F = pA \qquad (9.17)$$

where

  A = cross sectional area of rubber pad

  p = permitted specific pressure for rubber pad

According to Fig. 9.16, the cross-sectional area is $A = \dfrac{\pi}{4}(D^2 - d^2)$, for $f_{max} = 0.40\,h$ and rubber pad hardness 68 Shore, $p = 3.5$ MPa.

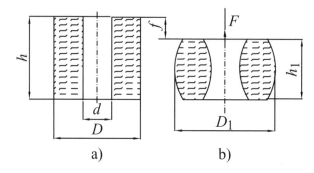

Fig. 9.16 Rubber pad: a) non-loaded; b) loaded.

9.5 DIE COMPONENTS FOR GUIDING AND STOPPING

The group of die components known as guides and stops includes the following components: stock guides, guide rails, French notch punches, pilots, and die stops.

9.5.1 Stock Guides and Guide Rails

a) Stock Guides

A good stock guide design always allows for staggering the entryway so that the workpiece will not snag. A good design also allows for the stock guide to be removable from the die without components having to be disassembled. Fig. 9.17 shows one possible design for a stock guide.

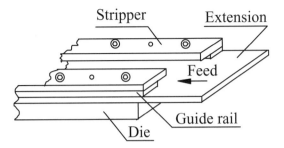

Fig. 9.17 Stock guide.

b) Guide Rails

Guide rails are used to guide the work strip through the die; they are placed between the stock shelf or die block and the stripper plate or guide plate.

The tunnel dimension A in Fig. 9.18 should allow free passage for the width of the stock and is calculated by the following formula:

$$A = B + 2c$$

where

B = strip width

c = clearance

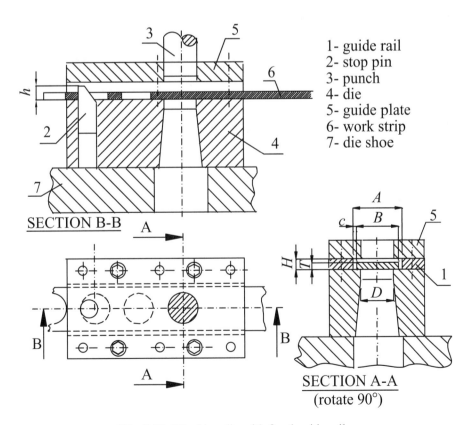

1- guide rail
2- stop pin
3- punch
4- die
5- guide plate
6- work strip
7- die shoe

SECTION B-B

SECTION A-A
(rotate 90°)

Fig. 9.18 Blanking die with fixed guide rails.

The tunnel height H in Fig. 9.18 provides clearance in the vertical direction. This dimension depends on the material thickness and method of work-strip feeds (manual or automatic). Advisable values for H and h are given in Table 9.4. Sometimes, to provide smooth movement to the workpiece in the production of large and complicated parts, four guide pins (two on each side) may be used instead of guide rails.

Table 9.4 Values for H and h

Material thickness T (mm)	Dimension H (mm)		Dimension h (mm)
	Manual feed	Automatic feed	
0.3 to 2.0	6.0 to 8.0	4.0 to 6.0	3.0
2.0 to 3.0	8.0 to 10.0	6.0 to 8.0	4.0
3.0 to 4.0	10.0 to 12.0	6.0 to 8.0	4.0
4.0 to 6.0	12.0 to 15.0	8.0 to 10.0	5.0
6.0 to 10.0	15.0 to 25.0	10.0 to 15.0	8.0

In compound dies, the elastic type of guide rails are used. Fig. 9.19 shows two types of design for elastic guide rails. These designs provide positive guidance of the work strip, and facilitate feeding through the die.

The clearance c for fixed guide rails is $c = 0.25$ to 0.75 mm. For the elastic-type guide rails (see Fig. 9.19), the clearance is $c = 2.5$ to 4.0 mm.

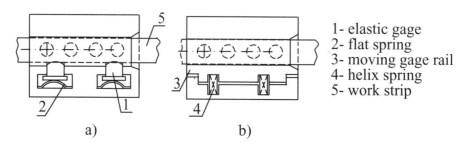

1- elastic gage
2- flat spring
3- moving gage rail
4- helix spring
5- work strip

a)   b)

Fig. 9.19 Elastic guides: a) gages with flat spring; b) gage with helix spring.

9.5.2 Die Stops and French Notch Punch

a) Pin Stop

The pin stops are used to stop the material strip after each feed movement is completed. Solid pins, with or without heads, may be used as stops. The pin should be lightly press-fitted into the die shoe and should extend above the die block face. The extension h of the die stops is a function of the thickness of the material. The value of h is given in Table 9.4. A clearance hole should be provided in the stripper plate and under the pin in the die shoe for removal of the pin when necessary.

b) French Notch Punch

The French notch punch is used for trimming away a length of work strip that is equal to the progression of the die. This action provides a fixed stop feature for strip progression such as is shown in Fig. 9.20.

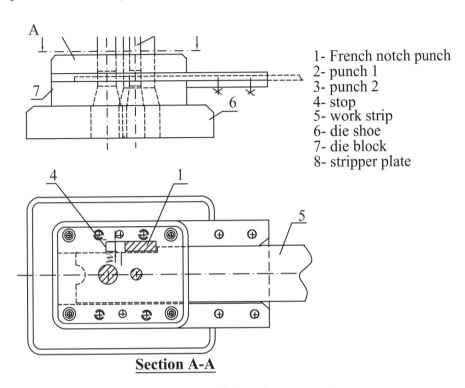

1- French notch punch
2- punch 1
3- punch 2
4- stop
5- work strip
6- die shoe
7- die block
8- stripper plate

Section A-A

Fig. 9.20 Die with French notch punch.

French notching is one of the best ways to control the problems of strip width tolerance, strip camber, and progression control. If French notching is used on only one side of the strip, a pilot punch must be used to allow the strip to back up 0.05 mm.

The value of the notch width (w) is a function of the thickness and kind of material. These values are given in Table 9.5.

Table 9.5 Values of the French notch width w

Material thickness T (mm)	Kind of material	
	Steel	Other softer materials
Up to 1.5	$(2 \text{ to } 3)T$	$(2 \text{ to } 3)T$
Over 1.5	$1.5T$	$2T$

Sometimes the back gage is extended and used to support a stop. This arrangement is especially useful with long workpieces. Such a stop is shown in Fig 9.21.

Finger stops are used to stop new strips in the proper location in a die. They are operated with the finger by pushing them into the stock channel until they seat.

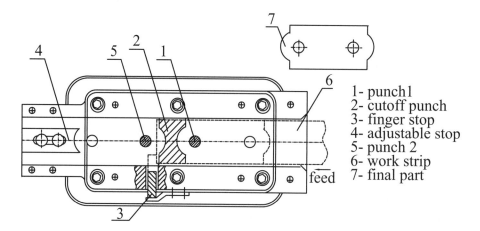

1- punch1
2- cutoff punch
3- finger stop
4- adjustable stop
5- punch 2
6- work strip
7- final part

Fig. 9.21 Compound die for punching holes and blanking long parts with finger stop and adjustable strip.

The press is tripped when the stop is released to its "out" position and it is not used again until a new strip is started. One type of finger stop is shown in Fig. 9.21. The die shown is sometimes used for punching holes and cutting workpieces.

9.5.3 Positioning the Individual Blank

There are many design options for positioning individual blanks or workpieces. The selection depends on the shape and dimension of the workpiece. Fig. 9.22 shows three methods of positioning the individual workpiece by using:

a) Three dowels

b) A ring

c) A combination of dowels and guide-rails

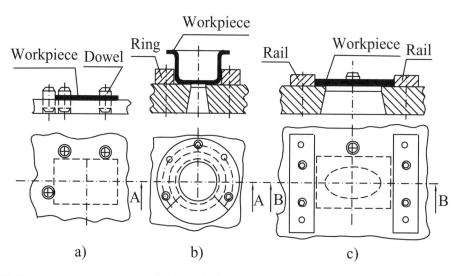

Fig. 9.22 Methods of positioning the individual blank or workpiece: a) with dowels; b) with ring; c) with rail.

9.5.4 Pilots

Pilots are used in progressive and compound dies to position the work strip so that the relationships between stations or previously punched holes and the outside blanked contours of work pieces may be maintained. Figure. 9.23 shows various methods for mounting pilots in punches.

The hole that receives the pilot should be extended through the punch so that the pilot may be removed if it breaks. The pilot should fit the work hole with a tolerance of from 0.02 to 0.15 mm. Pilots are generally made from tool steel, hardened and polished. They may have a spherical end that terminates in the diameter of the pilot, or the end of the pilot may be conical.

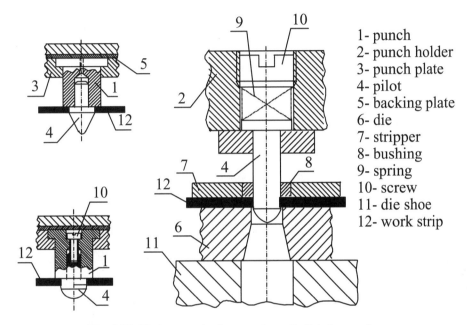

1- punch
2- punch holder
3- punch plate
4- pilot
5- backing plate
6- die
7- stripper
8- bushing
9- spring
10- screw
11- die shoe
12- work strip

Fig. 9.23 Various methods mounting of pilots in punches.

9.6 CENTER OF DIE PRESSURE

The die pressure should be centered on a vertical line passing through the specific point that defines the resultant force of the punching and blanking forces. There are two ways to determine the center of the die pressure: mathematically and graphically.

a) Mathematical Solution

As is known from statics, the coordinates of the point of resultant force are given by the formulas:

$$X = \frac{\sum_{i=1}^{i=n} F_i x_i}{\sum_{i-1}^{i=n} F_i}; \qquad Y = \frac{\sum_{i=1}^{i=n} F_i y_i}{\sum_{i=1}^{i=n} F_i} \qquad (9.18)$$

where

F_i = partial punch and blank forces

x_i ; y_i = coordinate center of gravity of partial punching and blanking parts

Assuming that the punch and blank forces are directly proportional to the length of the cut edges *Li*, the center of the die pressure may be determined by the following formulas:

$$X = \frac{\sum\limits_{i=1}^{i=n} L_i x_i}{\sum\limits_{i=1}^{i=n} L_i} = \frac{L_1 x_1 + L_2 x_2 + ... + L_n x_n}{L_1 + L_2 + ... + L_n}$$

$$Y = \frac{\sum\limits_{i=1}^{i=n} L_i y_i}{\sum\limits_{i=1}^{i=n} L_i} = \frac{L_1 y_1 + L_2 y_2 + ... + L_n y_n}{L_1 + L_2 + ... + L_n}$$

(9.19)

where

X; Y = coordinates of the center of the die pressure

x_i; y_i = coordinates of the center of gravity of a partial length of cut edge

L_i = partial length of cut edges

b) Graphical Solution

Subdivide the total length of the cut edges in Fig. 9.24 into partial lengths L_1, L_2, ...L_n, of which the center of gravity is known. The size of each partial length is represented as a force applied to the center of each partial length.

L_1- cut edge length of punch 1
$L_{2\&4}$- cut edge length of punch 2
L_3- cut edge length of punch 3
1- blanking punch
2- French notch punch
3- punching punch
4- work strip
S- center of die gravity

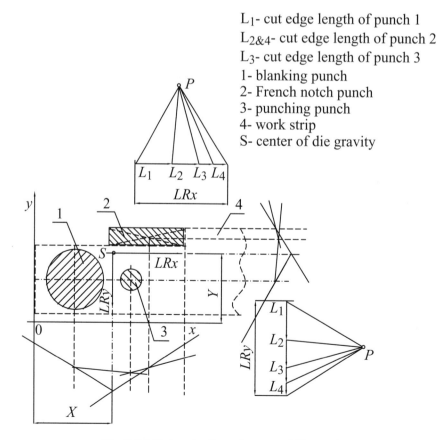

Fig. 9.24 Determination of center of die gravity.

Use the force polygon and link polygon closed to determine the main forces *LR*x and *LR*y operating in any two directions (preferably at right angles). The point of inter- section of the lines of application will indicate the position of the center of die pressure.

9.7 EXAMPLES OF CUTTING DIE DESIGNS

This section shows some examples of designs for punching, blanking, compound, and combination dies. Fig. 9.25 shows a design for a single blanking die. The die can produce either a right- or left-hand part.

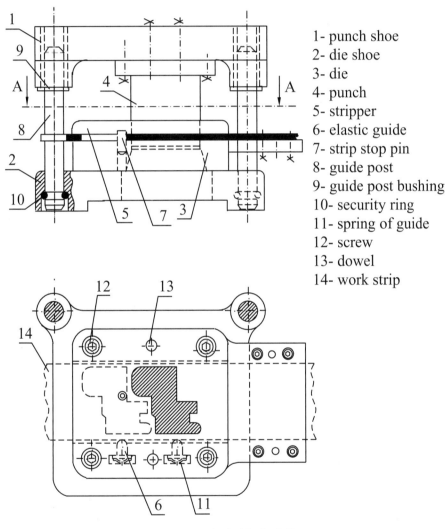

1- punch shoe
2- die shoe
3- die
4- punch
5- stripper
6- elastic guide
7- strip stop pin
8- guide post
9- guide post bushing
10- security ring
11- spring of guide
12- screw
13- dowel
14- work strip

Fig. 9.25 Single blanked die.

A flat blank, sheared by a blank-through type of blanking die, drops through the die block (lower shoe) and piles up on top of the bolster plate, or falls through a cored hole in the plate. The die consists of a punch shoe (1), and a die shoe (2), with a guide post (8), and a guide post bushing (9). The guide posts are press-fitted into the die shoe and secured. The blanking punch (4), is directly (i.e., without a punch plate) fixed to a punch holder with screws (12), and dowel (13). On the die shoe (2) is fixed the die block (3), the stripper (5), the guide rails (6), and the strip stop (7).

Figure 9.26 shows a punching die with a series of punches, which are staggered to reduce the force required to shear through metal, and to prevent punch breakage. The offset of the outside punches over the center punch is one full thickness of the material.

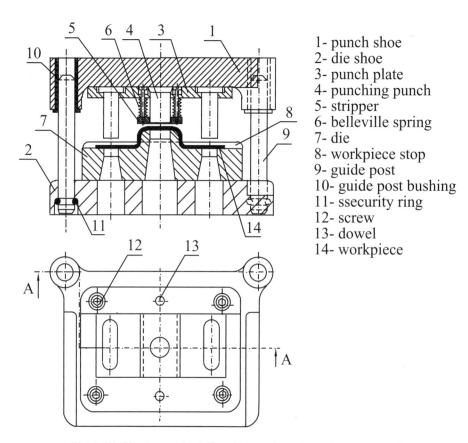

1- punch shoe
2- die shoe
3- punch plate
4- punching punch
5- stripper
6- belleville spring
7- die
8- workpiece stop
9- guide post
10- guide post bushing
11- ssecurity ring
12- screw
13- dowel
14- workpiece

Fig. 9.26 Single punched die with a series of punches staggered.

The die consists of a punch shoe (1) and die shoe (2), with guide post (9), and guide post bushing (10). The guide posts are press-fitted into the die shoe and secured. The punches (4) and the stripper (5), with the punch plate (3) and screws and dowels, are fixed to the punch holder.

The die shoe holds the die block (7), which also has the function of guiding the workpiece (14); the workpiece stop (8), which is attached to the die shoe with a screw (12); and dowels (13).

Fig. 9.27 shows a compound die with punching and blanking operations in parallel positions.

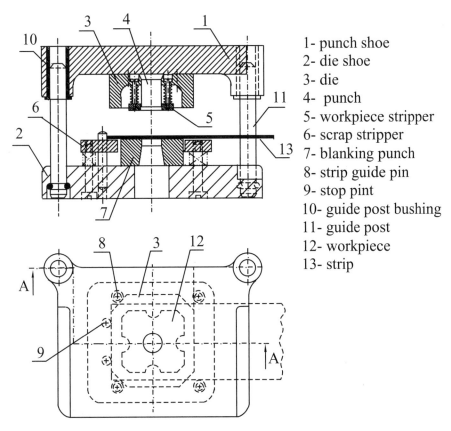

1- punch shoe
2- die shoe
3- die
4- punch
5- workpiece stripper
6- scrap stripper
7- blanking punch
8- strip guide pin
9- stop pint
10- guide post bushing
11- guide post
12- workpiece
13- strip

Fig. 9.27 Compound die with parallel positions for punching and blanking operations.

The die consists of a punch shoe (1) and a die shoe (2), with guide post (11) and guide post bushing (10). The guide posts are press-fitted into the die shoe and secured. The blanking die (3) is fitted on to the punch holder, along with the punching punch (4), and the workpiece stripper. On the lower shoe are fixed the blanking punch (7), and the scrap-strip stripper (6), the stop pin (9), and the strip guide-pins (8), which also function as guide bolts for the scrap-strip strippers.

Fig. 9.28 shows a compound die with progressively placed punching and blanking operations.

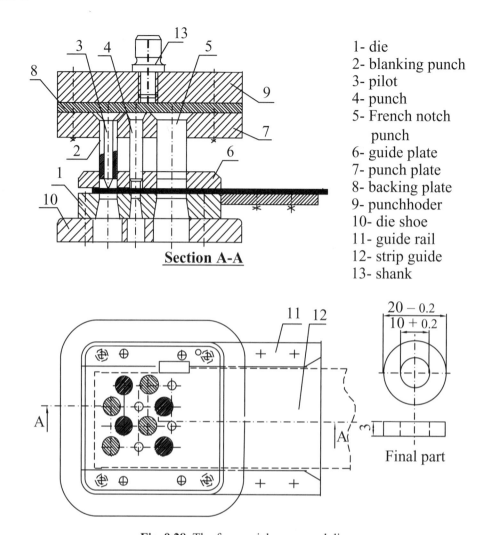

1- die
2- blanking punch
3- pilot
4- punch
5- French notch punch
6- guide plate
7- punch plate
8- backing plate
9- punchhoder
10- die shoe
11- guide rail
12- strip guide
13- shank

Section A-A

$20 - 0.2$
$10 + 0.2$

Final part

Fig. 9.28 The four-serial compound die.

On the punch holder (9) are mounted a punching punch (4), a blanking punch (2), and a French notch punch (5). Socket head screws and locating dowel pins are used to hold the punch plate (7) and the backing plate (8) to the punch holder.

The die holder (10) carries the die block, which consists of a die plate (1), a guide plate (6), and a strip guide (11). The punch shank (13) is shown because it is still in everyday use in many stamping shops. However, according to the OSHA Standard 1910.217(7) it cannot be used for clamping the punch holder to the slide (ram) of a press, but can be used for aligning the die in the press. Slide (ram) mounting holes or another clamping system must be provided in the punch holder for fastening.

The first punching holes are of 10 mm diameter; the blanking work piece is of 20 mm diameter. Pilots (3) are inserted into the blanking punches (2) to center the inner and outer contours.

The die is four-serial, so that every press stroke makes four pieces.

Ten

Bending Dies

10.1 INTRODUCTION

Bending dies may be conveniently classified according to their design, whether simple or complex, and according to their universality of application, whether universal or specif ic. The design of bending dies depends on the complexity of the workpiece shape, its dimensions, the type of material, the tolerances, etc.

10.2 SIMPLE DIE DESIGNS

Some die sets are designed to perform a single bending operation, which may include V, L, U, or Z bends and other profiles. Such dies are called single-operation dies or simple dies. One operation is accomplished with each stroke of the press.

10.2.1 U-Profile Bend Dies

Fig. 10.1 illustrates a simple die for bending a U-profile. In this example, the blank of length L, width b, and thickness T, is positioned on the die (1) between stop pins (9). The die (1) is mounted on the die holder (3) in the conventional position. The punch (2) is attached to the punch holder (4), which is fitted to the ram of the press. The pressure pad (8) applies pressure to the blank so that as the punch pushes the blank into the die, the workpiece is formed by a single stroke of the press. The bent workpiece is ejected from the die by the pressure pad mechanism (a) when the press ram retracts. For bending workpieces of small dimensions and thin material, the punches and dies are made from a single block of metal.

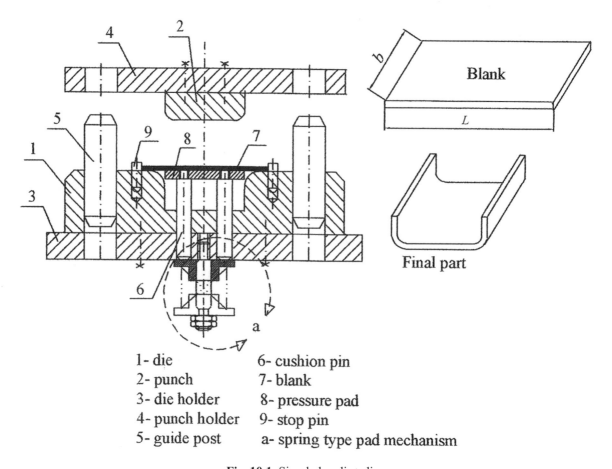

1- die
2- punch
3- die holder
4- punch holder
5- guide post
6- cushion pin
7- blank
8- pressure pad
9- stop pin
a- spring type pad mechanism

Fig. 10.1 Simple bending die.

For bending a larger workpiece, using a heavier die set (see Fig. 10.2), the die is made up of segments which in turn are generally made from alloy tool steel. The advantages of this design are the same as with punching and blanking dies, as described in Chapter 9.

The segments (4) are fixed to the die shoe with screws and dowels. The stop pin (7) is fixed to a pressure pad plate (6), which holds the blank in position while it is being worked on. The pressure plate also provides resistance to the bending operation, which is needed to perform the operation.

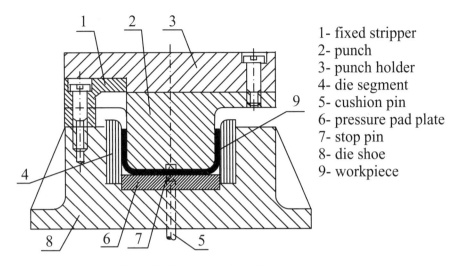

1- fixed stripper
2- punch
3- punch holder
4- die segment
5- cushion pin
6- pressure pad plate
7- stop pin
8- die shoe
9- workpiece

Fig. 10.2 Simple bending die with segments.

An air cushion or hydraulic cylinders are used to generate the force necessary for this resistance. If the width of the workpiece is large, it is necessary for the die to have a stripper. In the following description of a die set console, the stripper (1) is fixed to the die shoe, and the punch holder has a slot so that the upper set of the tool can move up and down without obstruction when the press ram cycles.

10.2.2 V-Profile Bend Dies

Fig. 10.3 shows a simple die set for bending different shapes with bending angles of ($0 < \alpha < 180°$). These types of die are known as dies for V-prof ile bends. The die (1) is mounted on the die shoe (4) in the conventional position. The punch (2) is attached to the punch holder (3). The pad (5) holds the blank in position while it is being worked on. It also provides the resistance needed to perform the bending operation. An air cushion or hydraulic cylinders are used to power this resistance. The lower and upper parts of the die set are guided by the guide post (7). The blank is positioned on the die (1), between the stop pins (6); with a single stroke of the press, the workpiece is bent into its required shape. Corrections to the punch angle, if necessary, are made after a try-out, when the exact springback angle is known.

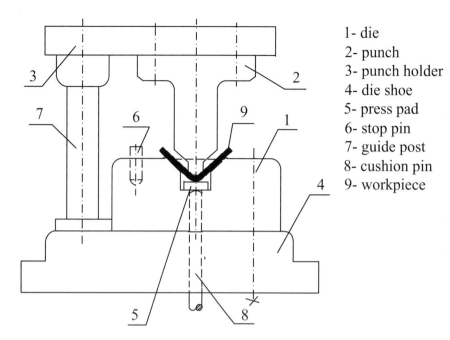

1- die
2- punch
3- punch holder
4- die shoe
5- press pad
6- stop pin
7- guide post
8- cushion pin
9- workpiece

Fig. 10.3 V-profile bend die.

Fig. 10.4 shows a single-operation bending die for a V-profile workpiece being bent on one end, with the other end of the piece already curled.

The die consists of a punch holder (3) and a die shoe (4) with guide posts (7). The guide posts are press fitted into the die shoe and secured. The bending punch (2) is fixed to the punch holder with screws and dowels. The die shoe carries the die (1), the blank support (9), and the stop pin (6). The pad (5) provides the resistance to the bending operation, needed to perform the operation. An air cushion or hydraulic cylinders are used to power this resistance.

The bent piece is ejected from the die with a pressure pad mechanism when the press ram is moved to the "up" position.

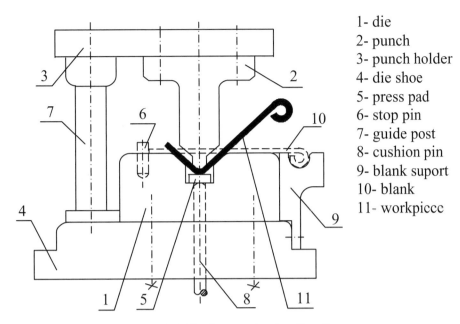

1- die
2- punch
3- punch holder
4- die shoe
5- press pad
6- stop pin
7- guide post
8- cushion pin
9- blank suport
10- blank
11- workpiccc

Fig. 10.4 Single operation bending die.

10.2.3 Universal Bending Dies

Figure 10.5 shows a typical simple universal die and an example of a workpiece bent in four phases. The die consists of a die shoe (4), on which are assembled the universal die (1) and adjustable stops (3). The die has several slots of different shapes. The punch (2) is attached directly to the press ram. Selecting an appropriate profile on the die (1) with an appropriate punch (2) allows the workpiece to be bent into different shapes as desired.

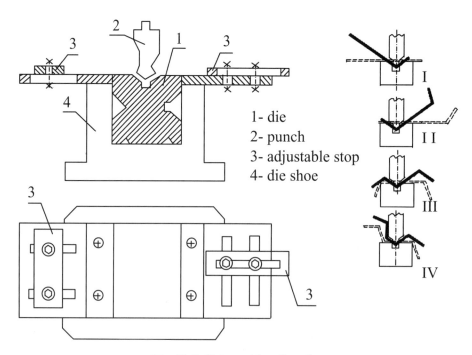

1- die
2- punch
3- adjustable stop
4- die shoe

Fig. 10.5 Universal bending die.

10.3 DIES OF COMPLEX DESIGN

"Complex" refers to dies that are made up of many elaborately interrelated or interconnected parts. Considerable study and knowledge are needed to design complex die systems.

10.3.1 Closing Profile Dies

A bent profile often is used as a starting point for the next bending operation to be performed on the final piece. The composite drawing in Fig. 10.6 illustrates both "before" and "after" stages of a bending die for closing a U-profile bend. The closing die consists of a punch holder (1) and a die shoe (2), which is guided by two guide posts and guide post bushings. The punch shoe (1) carries a holder (3) with movable side slides (4) that function as punches. Segments (6) are held in die shoe recesses with screws and dowels. The workpiece (U-profile, seen at the right, and in the die), with the insert (5) inside, is located on the central die segment. When the press ram is moved down, the side slides of the punch slide toward the die segment and with a single stroke of the press, the workpiece is bent to shape. The springs (7) return the slides to the start position when the ram press moves up.

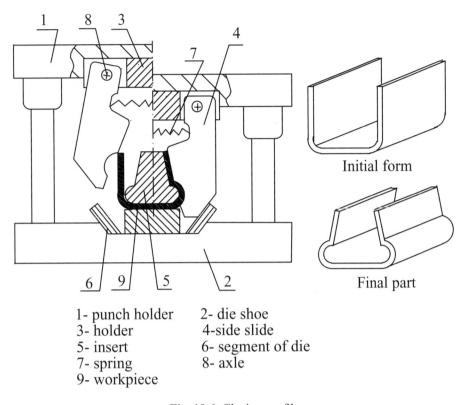

Fig. 10.6 Closing profile.

1- punch holder   2- die shoe
3- holder   4-side slide
5- insert   6- segment of die
7- spring   8- axle
9- workpiece

10.3.2 Special Bending Dies

If great precision is required in the bent pieces, a special design of die is used for bending and coining operations. One design of such a die is shown in Fig. 10.7. The die consists of a punch holder (1) and die shoe (2) with guide post and guide post bushing.

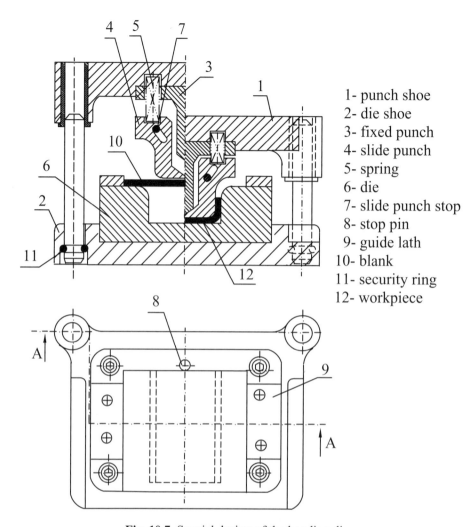

1- punch shoe
2- die shoe
3- fixed punch
4- slide punch
5- spring
6- die
7- slide punch stop
8- stop pin
9- guide lath
10- blank
11- security ring
12- workpiece

Fig. 10.7 Special design of the bending die.

The punch holder carries the punch, which consists of an immovable segment (3) and two movable segments (4), which are arranged to move transversely as the die closes. Attached to the die shoe are the die (6), with stop pin (8) and stop latch (9). After the bending operation is finished, the fixed segment of the punch (3) pushes the slide punches (4) so they move transversely and coin the outside of the workpiece.

If the inside dimensions of the workpiece need to be more precise, the punch is made of one piece, and the die is made of both movable and immovable segments. After the bending operation is finished, the punch additionally pressed to the movable segments of the die so they move transversely and coin the inside of the workpiece.

Figure 10.8 shows a die for bending a double closed L-profile. The die is shown in Fig. 10.8a; it includes the die shoe (8) and the punch (2), which is directly attached to the press ram by means of the shank (5). The die shoe embodies the die block, which consists of the immovable part of the die (1), movable segments (3), the workpiece stop (4), a pressure pad (9), pegs (6), and Belleville springs (7). The pressure pad holds the blank in position while it is being worked on; it provides the resistance to the bending operation needed to perform the operation. An air cushion or hydraulic cylinders are used to power this resistance.

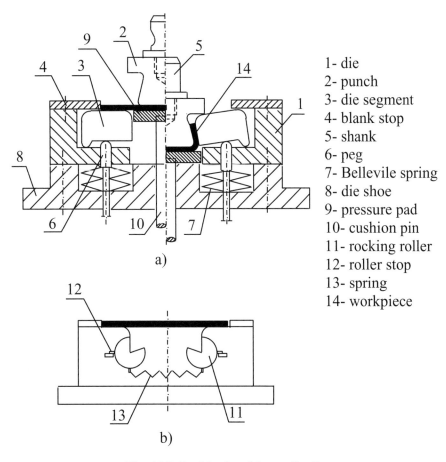

1- die
2- punch
3- die segment
4- blank stop
5- shank
6- peg
7- Bellevile spring
8- die shoe
9- pressure pad
10- cushion pin
11- rocking roller
12- roller stop
13- spring
14- workpiece

Fig. 10.8 Double closed L- profile die.

In the first bending phase, the punch (2) bends the blank to a double L-profile of 90 degrees. In the first bending phase, the pegs hold the movable segments of the die in the horizontal position. The Belleville springs are used to provide resistance until the blank has been bent to 90 degrees.

The second bending phase begins at the moment when the punch (2) touches the die segments (3). At this moment, the resistance of the Belleville springs (7) begins to be less than the force of the punch, the segments incline, and the workpiece is bent past 90 degrees. The stop (4) limits the inclination of the segments (3). When the press ram is moved up, the springs return the segments to the horizontal position, and the pressure pad ejects the workpiece.

Instead of the segments shown in Fig. 10.8a, roller cylinders may be used, as shown in Fig. 10.8b. This concept is simpler but has a weakness, which is the possibility of the blank slipping because it is not held during the bending operation.

10.3.3 Curling and Hinge Dies

Curling dies provide a curled or coiled-up end or edge to the piece. Hinge dies make use of a curling operation. The curl may be centered, or it may be tangential to the sheet, as shown in Fig. 10.9. The edge of the blank should have a starting bend and, if possible, the burr should be inside the bend. The blank is located on the lower shoe (12) where it is held by the pressure pad (6). The pressure pad, which holds the blank in position while it is being worked on, also provides resistance to sideways movement of the blank. The spring (7) is used to provide force to power this resistance. At the start of the operation, the bent edge of the blank is curled by a horizontally-moving cam slide die (1) that is forced inward by the cam driver (2). On the upstroke of the press, the spring (10) returns the sliding die to the starting position. The swallowtail die guide latch (11) provides a reliable guide for the cam slide die.

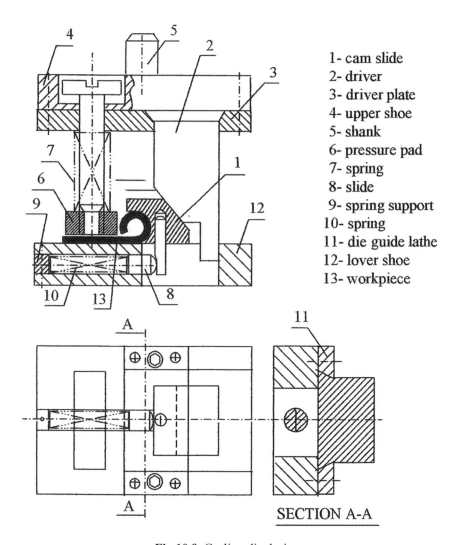

1- cam slide
2- driver
3- driver plate
4- upper shoe
5- shank
6- pressure pad
7- spring
8- slide
9- spring support
10- spring
11- die guide lathe
12- lover shoe
13- workpiece

SECTION A-A

Fig.10.9 Curling die design.

10.3.4 Tube-Forming Dies

Tube-forming dies may be single or multiple. Figure 10.10 shows a single die for forming a lube in three phases. The edges of the blank are bent in the first phase between the punch (1) and the die (2). The pressure pad (10) holds the blank in position while it is being worked on, and provides the resistance needed to perform the operation. The spring (11) provides resistance to the pressure pad. The workpiece is transferred by hand from the first position (phase one) to the second position (phase two). At the same time, a new blank is located at the first position. In the second phase, the workpiece is bent between the punch (3) and the die (4). The workpiece is then transferred to the third position, the second workpiece moves to the second position, a new blank is located at the first position, and so on. At phase three, the third position, forming of the workpiece is completed between the punch (5), the die (6), and the insert (7).

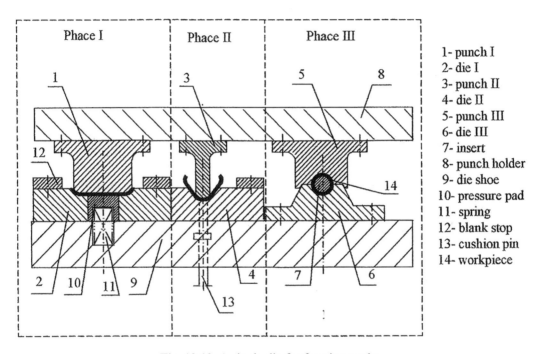

Fig. 10.10 A single die for forming a tube.

Even though each single press stroke makes one new piece after the completion of the first, productivity of this die design is low because the workpiece is transferred by hand. For mass production, the process needs to be automated. For production of more precise tubular pieces, a much better design for the die is shown in Fig. 10.11.

The tube-forming die in Fig. 10.11 consists of the punch holder (10) and a die shoe (11), with a guide system: guide post (8) and guide bushing (9). The punch (1) and driver (5) are attached to the punch holder. The die cam (4), the die segment (2), and the cam slide die (3), with springs (6) and (7), are fixed to the die shoe.

In operation, the blank (12) is located on the cam slide die. At the beginning of the bending operation, the blank is free bending. At the end of the downstroke of the press, the driver (5) pushes the die segments (3) to coin the workpiece around the punch (1) into its final dimensions. On the upstroke of the press, the springs (6 and 7) return the die segments to the start point and the tube is slipped off the punch (1).

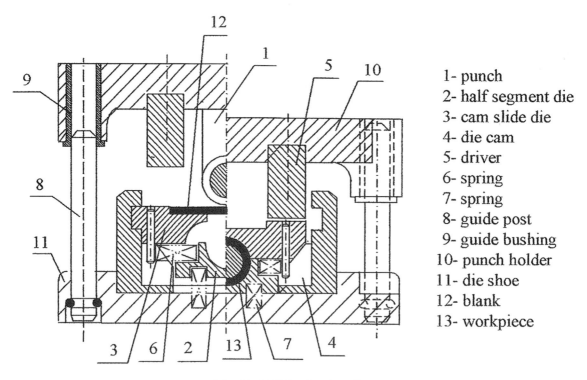

1- punch
2- half segment die
3- cam slide die
4- die cam
5- driver
6- spring
7- spring
8- guide post
9- guide bushing
10- punch holder
11- die shoe
12- blank
13- workpiece

Fig. 10.11 Tube-forming die.

10.3.5 Multiple-Bend Dies

Multiple-bend dies offer an infinite variety of possibilities. Commonly used in mass production, these dies can accomplish, in a single stroke, an operation that would require several operations with a single-bend die. Such a die requires much greater pressures than those required by dies for individual operations, and an example is shown in Fig. 10.12.

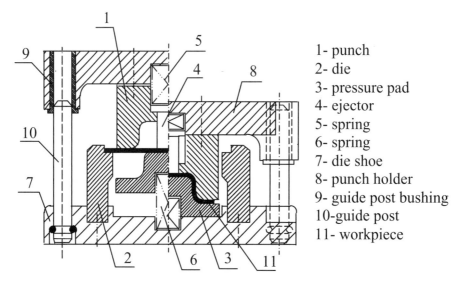

1- punch
2- die
3- pressure pad
4- ejector
5- spring
6- spring
7- die shoe
8- punch holder
9- guide post bushing
10- guide post
11- workpiece

Fig.10.12 Multiple-bend die.

Attached to the punch holder (9) is the punch (1), whose inside surface has the function of a die during the final bending phase. The die shoe carries the die (2) and the pressure pad (3), with the spring (6). The pressure pad functions as a punch during the final bending phase. The first blank is bent into a U-profile by the action of the outside of the punch (1) and the die (2). Then the pressure pad (3) and the inside of the punch (1) act to form the final shape of the piece. The piece is removed by means of the ejector (4) and the spring (5), located in the upper set of the die.

10.3.6 Combination Dies

A combination die is a simple-station complex die in which both cutting and non-cutting operations are accomplished at one press stroke. An example of such a die is shown in Fig. 10.13.

The die in Fig. 10.13 consists of the punch holder (1), the die shoe (2), with guide system/guide post (10), and the guide post bushing (11). The punch holder carries the bending punch (3), which functions as a punch plate for the punch (4). The punch (4), bears the pressure pad (6), and the spring (15) is also attached to the punch holder as well as the strip cutting punch (13).

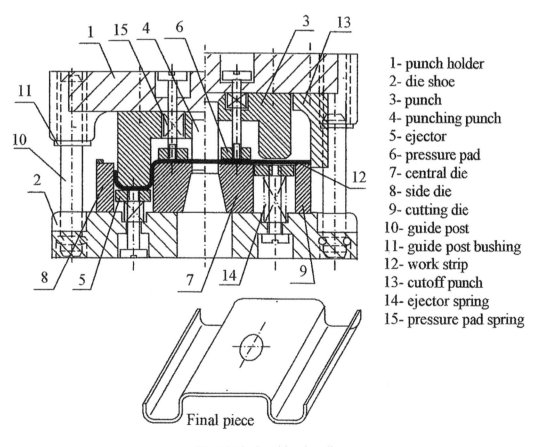

1- punch holder
2- die shoe
3- punch
4- punching punch
5- ejector
6- pressure pad
7- central die
8- side die
9- cutting die
10- guide post
11- guide post bushing
12- work strip
13- cutoff punch
14- ejector spring
15- pressure pad spring

Fig.10.13 Combination die.

The die shoe carries a central die (7) for bending and punching operations; a side die (8) for bending one end of the workpiece; and a side die (9) for cutting the work strip (12) and bending the other end of the workpiece. The die shoe also supports the workpiece ejector (5) and spring (14).

In the first operating phase of the combination die in Fig. 10.13, the cutting punch (13) cuts the work strip, and the punch (4) then punches a hole. As the punch continues to descend, the bending punch (3), the central die (7), and the side dies (8 and 9), bend the workpiece into its final shape.

10.3.7 Progressive Dies

Individual operations in a progressive die are often relatively simple, but when they are combined into several stations it is often difficult to devise the most practical and economical strip design for optimum operation of the die.

In designing a die to produce good pieces, the sequence of operations for a strip and the details of each operation must be carefully developed. A tentative sequence of operations should be established and the following items should be considered as the final sequence of operations is developed:

1. Arrange for piloting holes and piloting notches to be punched in the first station. Other holes may be punched that will not be affected by subsequent non-cutting operations.
2. Develop the blank for drawing or forming operations for free movement of strip.
3. Distribute punching areas over several stations if they are close to each other or close to the edge of the die opening.
4. Analyze the shapes of blanked areas in the strip for division into simple shapes, so that commercially available punches may be used for simple contours.
5. Use idle stations to strengthen die blocks and stripper plates, and to facilitate strip movement.
6. Determine if strip grain direction will adversely affect or facilitate any operation.
7. Plan the bending or drawing operations in either an upward or a downward direction, whichever will assure the best die design and strip movement.
8. The shape of the finished piece may dictate that the cutoff operation should precede the last non-cutting operation.
9. Design adequate carrier strips or tabs.
10. Check strip layout for minimum scrap. Use a multiple layout, if feasible.
11. Locate cutting and forming areas to provide uniform loading of the press slide.
12. Design the strip so that the scrap and part can be ejected without interference.

Figure 10.14 shows a progressive die with four workstations. In this die, after four press strokes down, every following press stroke down makes one final piece. The die block is machined from four pieces and fastened to the die shoe. This arrangement permits the replacement of broken or worn die blocks.

The stock is fed from the right. The first strip is stopped by a finger stop (not shown). The first down stroke of the press (Fig. 10.14-I) produces a bigger hole and two notches. These notches form the left end of the first piece.

The press ram retracts, and the stock moves to the next station. The second station is idle (Fig. 10.14-II). The right end of the second piece and a second bigger hole have now been punched.

The press ram retracts a second time, and the scrap moves to the third station (Fig. 10.14-III). The third ram stroke punches the four small holes as shown in Fig. 10.14-III. The fourth ram stroke (Fig. 10.14-IV) bends the sides, and cuts off and forms the end radii of the finished piece. Thereafter, every press down stroke produces a finished piece, as shown in Fig. 10.14-V.

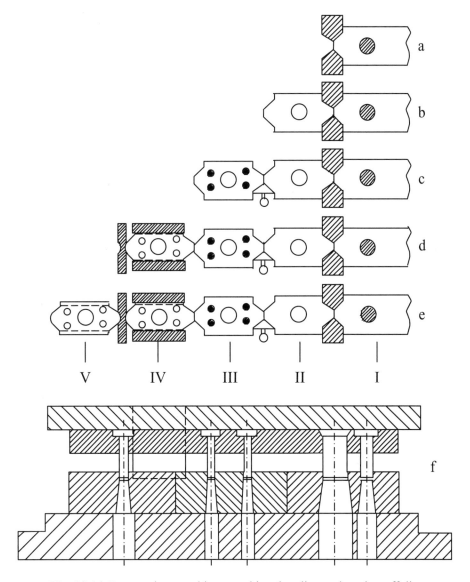

Fig. 10.14 Progressive punching, notching, bending and cutting-off die.

Eleven

Deep Drawing Dies

11.1 INTRODUCTION

The important variables in the technical components of deep drawing dies are the punch corner radius, the die ring profile, the clearance between punch and die ring, and the configuration of the surfaces of the die rings and the blank holder that are in contact with the blank.

11.2 DRAW RINGS

The die ring profile substantially influences both the deep drawing process and the quality of the drawn workpieces. Because the material is pulled over the profile, it is necessary that the die ring profile have an optimum value. The most frequently used draw rings use a corner radius or conic profile, even though other kinds of draw rings are used as well.

11.2.1 Draw Ring with Corner Radius

A draw ring with a corner radius (R_p) is the most frequently used for the first drawing operation without a reduction of the thickness of the materials, whether a blank holder is used or not. Figure 11.1 shows a deep drawing die with a draw ring having a corner radius.

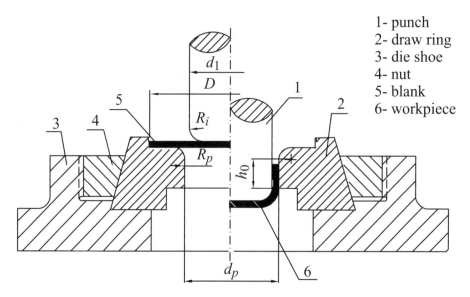

1- punch
2- draw ring
3- die shoe
4- nut
5- blank
6- workpiece

Fig. 11.1 Die for the first drawing operation: die ring with corner radius.

A draw ring with a corner radius, as shown in Fig. 11.2, can also be used for subsequent drawing operations without any reduction in the thickness of the materials if a blank holder is not used. If a blank holder is used, however, the conditions for drawing are much better using a draw ring with a conic profile than one with a corner radius.

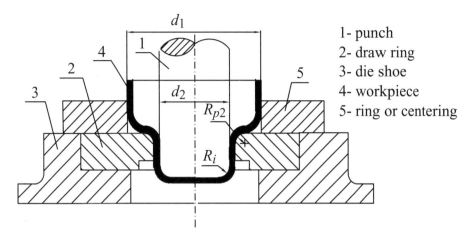

1- punch
2- draw ring
3- die shoe
4- workpiece
5- ring or centering

Fig. 11.2 Typical die for subsequent drawing operations using a draw ring with corner radius.

a) Cylindrical Shells

The value for the optimum radius on the draw ring is defined experimentally, and it depends on the type and thickness of material being drawn, the order of the drawing operation, the height of the workpiece, and the drawing ratio.

The smaller the draw ring corner radius, the greater the force needed to draw the shell. If the corner radius of the draw ring is too large, too much of the material will not be confined as it passes over the radius. The material will thicken, fold, and wrinkle. Recommendations for approximating the draw ring radius can be found in the technical literature. E. Kaczmarek recommends the following formula for defining the draw ring corner radius for the first drawing operation:

$$R_{\mathrm{p}} = 0.8 \cdot \sqrt{(D - d_1)T} \qquad (11.1)$$

where

D = blank diameter

d_1 = inside workpiece diameter after the first drawing operation

T = material thickness

The draw ring corner radius for the next drawing operation is:

$$R_{\mathrm{p(n)}} = 0.8 \cdot \sqrt{(d_1 - d_n)T} \qquad (11.1a)$$

where

d_{n} = inside shell diameter after n-th drawing operation

The height of the cylindrical part of the draw ring (h_0) in Fig. 11.1 can be calculated by the following formula:

$$h_0 = (3 \text{ to } 5)T \qquad (11.2)$$

b) Noncylindrical Shells

The die draw radius (R_{p}) for drawing a rectangular or square shell is given by the following formulas:

For longer side (a)

$$R_{p(a)} = 0.035\left[50 + 2\left(L_a - a_1\right)\right]\sqrt{T} \tag{11.3}$$

For shorter side (b)

$$R_{p(b)} = 0.35\left[50 + 2\left(L_b - b_1\right)\right]\sqrt{T} \tag{11.3a}$$

For a corner radius (Re), the draw radius is:

$$R_{p(e)} = 2.5R_{p(a)} \tag{11.3b}$$

where

L_a, L_b = blank dimensions

a_1, b_1 = shell dimensions after the first drawing operation

T = material thickness

11.2.2 Draw Ring with Conical Profile

A draw ring with a conical profile, as shown in Fig. 11.3, is used for the first drawing operation only if the blank is drawn without a blank holder. However, a draw ring with a conical profile is frequently used for later drawing operations with a reduction of wall thickness, as well as for subsequent drawing operations without a reduction in the wall thickness of the drawing shell, because such a profile is better able to hold the workpiece.

The central angle of the cone on the draw ring for deep drawing without a reduction in workpiece thickness is:

$$\alpha = 40° \text{ to } 45°, \text{ sometimes to } 52°$$

For drawing with a reduction in the thickness of the workpiece wall, the angle is:

$$\alpha = 12° \text{ to } 18°$$

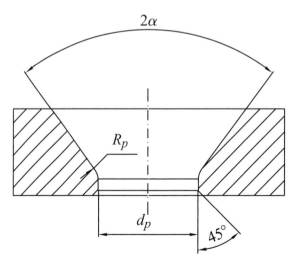

Fig. 11.3 Draw ring with conical profile.

To perform deformations by diameter (bending deformation) separately from deformations by wall material thickness (drawing deformation), for drawing operations with a reduction of the wall thickness of the shell, the draw ring profile often combines both radius and conic shapes. Figure 11.4 shows this type of draw ring. Such a draw ring may be solid, as shown in Fig. 11.4a, so that the deformations accomplished by reducing the diameter are performed by a draw ring with a corner radius. Draw rings may also be made as a combination of two rings, as in Fig. 11.4b, so that deformations that result in a change in the thickness of the material are performed by a draw ring with a conic profile.

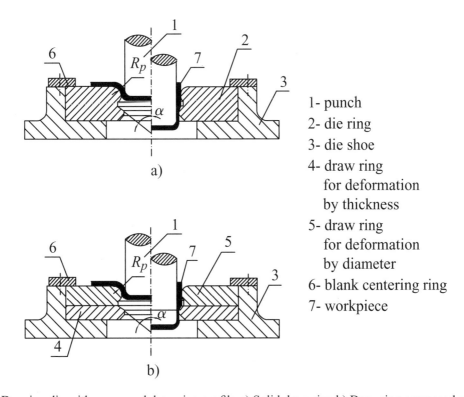

1- punch
2- die ring
3- die shoe
4- draw ring
    for deformation
    by thickness
5- draw ring
    for deformation
    by diameter
6- blank centering ring
7- workpiece

Fig. 11.4 Drawing die with composed draw ring profile a) Solid draw ring b) Draw ring composed of two rings.

Using a combination of two draw rings works better because separate draw rings are easier to make and easier to replace if worn or broken. The draw ring (5), Fig 11.4b, used for deformations by diameter, and the draw ring (4), Fig. 11.4b, used for deformations involving wall thickness. are assembled into a die shoe (3) and fixed with screws and a blank centering plate. This combination of draw rings enables good drawings to be made, as opposed to those achieved with draw rings having a corner radius. The difference occurs because the maximum forces of deformation by diameter and deformation by wall thickness of the workpiece do not occur at exactly the same time and therefore do not apply excessive forces.

A draw ring with only a corner radius may be chosen to perform simultaneous deformation by diameter and reduction in the thickness of the workpiece in the first drawing operation. If this procedure is selected, the maximum forces of both deformation by diameter and deformation by thickness will be superposed and may result in cracks in the workpiece.

11.3 CLEARANCES AND TOLERANCES

The clearance between the walls of the punch and the die is a very important parameter in deep drawing operations. In a drawing operation with no reduction in workpiece thickness, the material clearance should be greater than the thickness of the material. If the clearance is too small, the blank may be pierced or sheared by the punch. The clearance value may be defined either as a percentage of material thickness or by an empirical formula depending on the kind and thickness of the material.

$$c = T + k\sqrt{10T} \qquad (11.4)$$

where

c = clearance

T = material thickness

k = coefficient (Table 11.1)

Table 11.1 Values of coefficient k for different materials

Material	Coefficient k
Steel sheet	0.07
Aluminum sheet	0.02
Other metal sheet	0.04

11.3.1 Calculation of the Dimensions of the Punch and Die

When the value of the clearance c is known, calculation of the punch dimensions can be according to whether it is the outside or the inside dimension of the final piece that must be within a given tolerance. The calculations will be different for the inside and outside dimensions:

    a) If the outside diameter of the final piece must be within a certain tolerance, the draw ring diameter (d_p) is equal to the minimum outside diameter of the final piece, and the punch diameter is less than the draw ring diameter by $2c$. The nominal draw ring diameter is:

$$d_p = d_0 - \Delta \qquad (11.5)$$

    The draw ring and punch are assigned working tolerances (t_p, t_i), given in Table 11.2, where d is the nominal diameter of the final piece and T is the thickness of the work material.

Table 11.2 Work tolerances for draw ring and punch

d (mm)	Tol.	Material Thickness T (mm)									
		0.25	0.35	0.50	0.60	0.80	1.0	1.2	1.5	2.0	2.5
10 to 50	$+t_p$	0.02	0.03	0.04	0.05	0.07	0.08	0.09	0.11	0.13	0.15
	$-t_i$	0.01	0.02	0.03	0.03	0.04	0.05	0.06	0.07	0.08	0.10
51 to 200	$+t_p$	0.03	0.04	0.05	0.06	0.08	0.09	0.10	0.12	0.15	0.18
	$-t_i$	0.01	0.02	0.00	0.04	0.05	0.06	0.07	0.08	0.10	0.12
201 to 500	$+t_p$	0.03	0.04	0.05	0.06	0.08	0.10	0.12	0.14	0.17	0.20
	$-t_i$	0.01	0.02	0.03	0.04	0.06	0.07	0.08	0.09	0.12	0.14

The maximum draw ring diameter is:

$$d_{p(max)} = d_p + t_p = d_0 - \Delta + t_p \tag{11.5a}$$

The nominal punch diameter is:

$$d_i = d_p - 2c = d_0 - \Delta - 2c \tag{11.6}$$

The minimum punch diameter is:

$$d_{i\,(min)} = d_i - t_i = d_0 - \Delta - 2c - t_i \tag{11.6a}$$

where

   d_p, d_i = nominal draw ring and punch diameter

   d_0 = nominal diameter of the outside of final piece

   Δ = final piece's working tolerance

   t_p, t_i = work tolerance of draw ring and punch

   c = clearance

b) If the inside diameter of the final piece is within the tolerance, the punch diameter (d_i) is equal to the minimal inside diameter of the final piece, and the draw ring diameter is larger than the punch diameter for 2c.

The nominal punch diameter is:

$$d_i = d_u \tag{11.7}$$

The minimal punch diameter is:

$$d_{i\,(min)} = d_i - t_i = d_u - t_i \tag{11.7a}$$

The nominal draw ring diameter is:

$$d_p = d_i + 2c = d_u + 2c \tag{11.8}$$

The maximum draw ring diameter is:

$$d_{p\,(max)} = d_p + t_p = d_u + 2c + t_p \tag{11.8a}$$

where

d_u = nominal diameter of the inside of the final piece

It is very important that the clearance be kept constant during deep drawing operations; otherwise, the final piece may have unequal wall thickness. The best solution to centering the punch and draw ring are dies with a guide post system, which is better than other solutions such as dies with self-centering elastic rings.

c) The punch nose radius for deep drawing operations without wall thickness reduction may be calculated by the following formula:

$$R_{i(i)} = \frac{d_i - d_{i-1}}{2} \tag{11.9}$$

d) Sometimes, instead of a punch nose radius, a conic-shaped punch nose may be used with a bevel angle $\beta = 45°$ to $50°$ and transfer radius:

$$R_{i\,(i)} = R_{p\,(i+1)} \tag{11.9a}$$

where

d_i, $d_{(i-1)}$ = punch diameter i-th and (i − 1) drawing operation

$R_p(i + 1)$ = draw ring radius (i + 1) drawing operation

β = conic bevel angle

A punch with a conic nose is used in combination with a conic draw ring. The punch must have a ventilation channel to avoid deformation of the bottom of the workpiece.

e) The punch profile for drawing operations with wall thickness reduction is shown in Fig. 11.5. The punch is made with two conic shapes. The first cone has a bevel angle $\beta_1 = 2°$ to $4°$ for the first drawing operation, and for the next drawing operations, this angle is gradually reduced depending on the workpiece form.

The second cone follows the first cone and has a bevel angle $\alpha_1 \le 1°$ for the first operation. For the following operations, this angle is gradually reduced until the final inside dimension of the workpiece is attained. The second cone is used to ease removal of the workpiece from the punch.

The punch nose radius depends on the thickness and type of workpiece material. For the first drawing operation the punch nose radius is:

$$R_{i(1)} = (0.9 \text{ to } 1.0)T \quad \text{for brass} \tag{11.9b}$$
$$R_{i(1)} = (1.3 \text{ to } 1.4)T \quad \text{for steel.}$$

For the following operations, this value is gradually reduced until the final inside dimension of the workpiece is reached

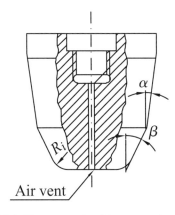

Fig. 11.5 Draw punch with two conic angles.

11.4 BLANK HOLDERS

The blank holder's function is to prevent the appearance of wrinkles in the top flange of the shell. The most important parameter is the blank holder pressure. If the blank holder exerts too little pressure, or if the punch or draw ring radii are too large, wrinkles will appear. Wrinkles are sometimes caused by too much metal trying to crowd over the draw ring, which usually happens when the clearance between the punch and the die is not great enough.

To determine whether or not it is necessary to use a blank holder, consider:

- the ratio of the relative material thickness (T_r)
- the drawing ratio (m)

A shell can be drawn without a blank holder if the following conditions are satisfied:

1) *Cylindrical shell*

    For the first operation:

$$T_r = \frac{T}{D}100 \geq 2\% \quad \text{and} \quad m_1 = \frac{d_1}{D} \geq 0.6 \tag{11.10}$$

    For subsequent operations:

$$T_r = \frac{T}{d_{i-1}}100 \geq 1.5\% \quad \text{and} \quad m_i = \frac{d_i}{d_{i-1}} \geq 0.8 \tag{11.10a}$$

2) *Hemisphere*

    For all operations:

$$T_r = \frac{T}{D}100 \geq 3\% \tag{11.11}$$

where

D = blank diameter

d_1, d_i, d_{i-1} = cup diameter after the first, i, and (i − 1) operation

T = material thickness

If these conditions are not satisfied, a blank holder is necessary.

11.4.1 Blank Holder Pressure

The blank holder pressure can be calculated by the following formulas:

    a) For the first drawing operation:

$$p_{d1} = \left(0.2 \text{ to } 0.3\right)\left[\left(\frac{D}{d_1}-1\right)^3 + \frac{d_1}{200T}\right]\cdot\left(UTS\right) \tag{11.12}$$

    where

        D = blank diameter

        T = material thickness

        d_1 = inside cup diameter after the first drawing operation

        UTS = ultimate tensile stress of material

    b) For subsequent drawing operations:

$$p_{di} = \left(0.2 \text{ to } 0.3\right)\left[\left(\frac{d_{i-1}}{d_i}-1\right)^3 + \frac{d_i}{200T}\right]\cdot\left(UTS\right) \tag{11.12a}$$

11.4.2 Blank Holder Force

The blank holder force can be calculated by the following formula:

    a) For the first drawing operations:

$$F_{d1} = \frac{\pi}{4}\left[D^2 - \left(d_{p1}+2R_{p1}\right)^2\right]p_{d1} \tag{11.13}$$

    b) For subsequent operations (see Fig. 11.6)

$$F_{di} = \frac{\pi}{4}\left[d_{i-1} - \left(d_1+2T\right)^2\right]\cdot\left(1+\mu\mathrm{ctg}\alpha\right)p_{d1} \tag{11.13a}$$

    where

        α = die ring conic bevel angle

        μ = coefficient of friction (μ = 0.1 to 0.15).

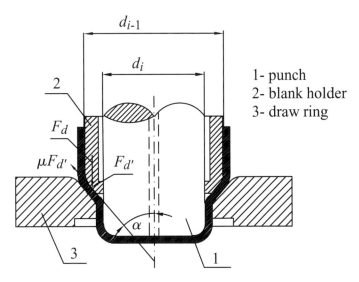

Fig. 11.6 Schematic diagram of blank holder for subsequent drawing operations.

For a die with a large draw ring radius, for example ($R_p = 20T$), and a wide punch nose radius ($R_i = 25T$), it is necessary to use two blank holders, as shown in Fig. 11.7.

The type of die in Fig 11.7 is designed for use on a double-action press. At first, the press ram moves the blank holder (4) down, putting pressure on the blank so that the punch (1) pushes the blank through the draw ring (2). The driver (7), which is attached to the punch and fixed with a nut (8), moves down at the same time, putting pressure on the rubber ring (6) and the workpiece holder (5). The workpiece is held with sufficient force to allow the material to slide from under the holder and move over the draw ring radius.

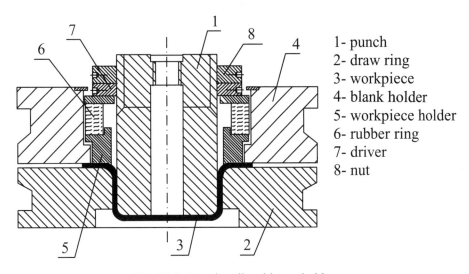

Fig. 11.7 Drawing die with two holders.

11.4.3 Draw Beads

Draw beads are often necessary to control the flow of the blank into the die ring. Beads help or restrict the flow of the material by bending or unbending it during drawing.

The draw beads that help the flow of the material are located on the die ring. The dimensions of these draw beads are shown in Fig. 11.8.

Radius R can be calculated by the following formula:

$$R = 0.05 d_i \sqrt{T} \tag{11.14}$$

where

T = material thickness

d_i = punch diameter

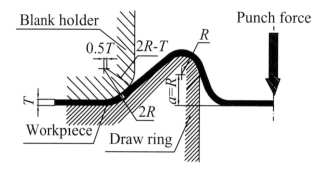

Fig.11.8 Draw bead for helping flow of the material.

Draw beads for restricting the flow of material are located on the blank holder and in corresponding places on the die ring there are slots. Figure 11.9 shows the bead design, their arrangement around the workpiece's drawing contour, and recommended dimensions. These draw beads are especially necessary in drawing rectangular shells and nonsymmetrical pieces.

Draw beads also help reduce required blank holder forces.

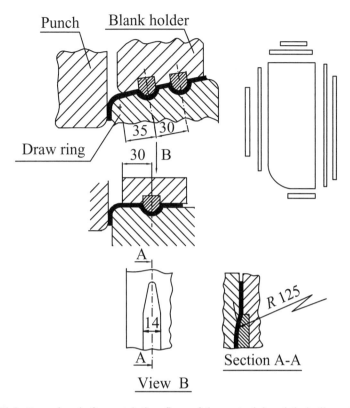

Fig.11.9 Draw beads for restricting flow of the material and their disposition.

11.5 SINGLE-OPERATION DIES

These types of dies can be designed for single or for double-action presses. In the first case, the blank holder gets its power from a mechanism located below the bed of the press.

Fig. 11.10 shows a die for a drawing operation on a single-action press. The die in Fig. 11.10 consists of the upper shoe (2) and lower shoe (9), with guide posts (7). The die ring (3) and ejector (5) are attached to the upper shoe. The punch (4), the pressure pad (6), and the cushion pin (8) are attached to the lower shoe. A blank is inserted into blank stop ring (1), which is fixed on the pressure pad (6). The punch is vented to aid in stripping the workpiece from the punch.

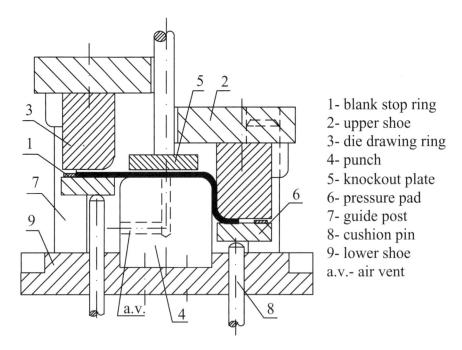

1- blank stop ring
2- upper shoe
3- die drawing ring
4- punch
5- knockout plate
6- pressure pad
7- guide post
8- cushion pin
9- lower shoe
a.v.- air vent

Fig. 11.10 Single-operation drawing die for single-action press.

Figure 11.11 shows a single-operation drawing die of similar conception, but with a different workpiece pressure pad. The die consists of the upper shoe (5) and the lower shoe (7), with guide posts (8) and (10), and guide post bushings (9). To the upper shoe are attached the die ring (2), held by the die ring holder (6) and the workpiece ejector (4). To the lower shoe are attached the punch (3) and the pressure ring (1), powered by a mechanism located below the bed of the press. This kind of die is used for the second and later drawing operations.

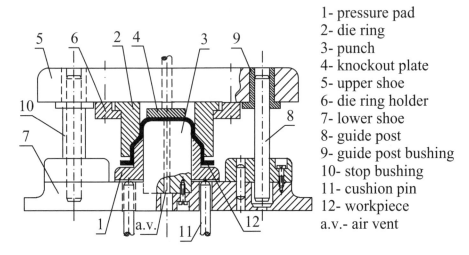

1- pressure pad
2- die ring
3- punch
4- knockout plate
5- upper shoe
6- die ring holder
7- lower shoe
8- guide post
9- guide post bushing
10- stop bushing
11- cushion pin
12- workpiece
a.v.- air vent

Fig. 11.11 Single-operation drawing die.

A single-operation die for the first drawing operation on a double-action press is shown in Fig.11.12. The die consists of the upper shoe (6) and lower shoe (7). To the upper shoe is attached a blank holder (1). To the lower shoe are fixed the die ring (3); elastic blank stop pins (5), inserted into the die ring; and the workpiece ejector

(4). The punch is attached directly to the inner slide, and the blank holder to the outer slide. The pressure pad puts sufficient pressure on the blank so that as the punch pushes the blank through the draw ring, the workpiece is held with a force great enough to prevent the material from rising and light enough so that the material is able to slide from under the pad and move over the die ring radius. The workpiece ejector is powered by a mechanism located below the bed of the press.

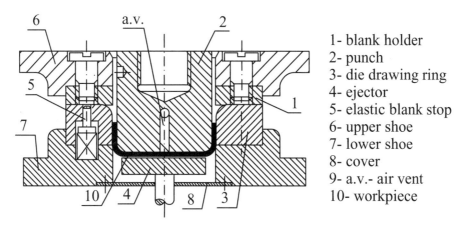

1- blank holder
2- punch
3- die drawing ring
4- ejector
5- elastic blank stop
6- upper shoe
7- lower shoe
8- cover
9- a.v.- air vent
10- workpiece

Fig. 11.12 Single-operation die for double-action press.

11.6 MULTI-OPERATION DIES

Multi-operation dies are combination dies designed to perform both drawing and non-drawing operations in one press stroke. Figure 11.13 shows a multi-operation die intended for use in a single-action press. First, a blank is cut from a strip of stock with a blanking punch (2), and blanking die (1). Then, the drawing operation is performed with the drawing punch (3) and the inside of the blanking punch (2), which functions as the draw ring.

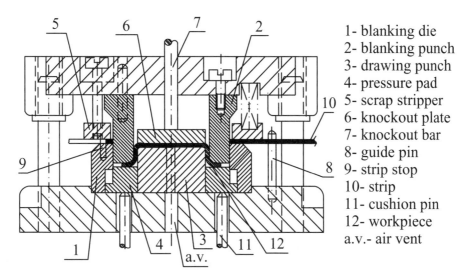

1- blanking die
2- blanking punch
3- drawing punch
4- pressure pad
5- scrap stripper
6- knockout plate
7- knockout bar
8- guide pin
9- strip stop
10- strip
11- cushion pin
12- workpiece
a.v.- air vent

Fig. 11.13 Multi-operation die for single-action press.

The pressure pad (4) puts pressure on the blank so that as it is pushed through the drawing ring, the blank is held in place. Enough force is exerted to prevent the material from rising, but not enough to prevent the material from being able to slide from under the pad and over the die ring radius. When the press slide moves up, the stripper pulls scrap strip from the blanking punch, and the knockout bar pushes the workpiece out. A strip of stock is guided by the guide pins (8) and stopped by a stop pin (9). The pressure pad gets power from a mechanism located under the press bed.

11.7 PROGRESSIVE DRAWING DIES

Progressive drawing dies are designed to do separate operations: blanking, punching (if necessary), drawing, and redrawing shells in successive workstations. These dies are used to produce pieces of smaller dimensions in mass production. A strip of stock is usually moved automatically through the die. Moving the drawn shell from one station to the next through progressive dies is sometimes difficult, especially if the shell is drawn deeply. There are two drawing methods: in the first, the workpiece is not initially cut so that it may be carried from one station to the next until it reaches the last station, where it is cut and pushed out of the die. In the second method, the blank is partially retained in the strip (called "cut and carry") and carried to the scrap strip from one station to the next until the workpiece reaches the last station, where it is finally cut clear and pushed out of the die.

Figure 11.14 shows a progressive die for the drawing of a bushing by the first method, and the scrap strip with successive shells drawn in this die (Fig.11.14a).

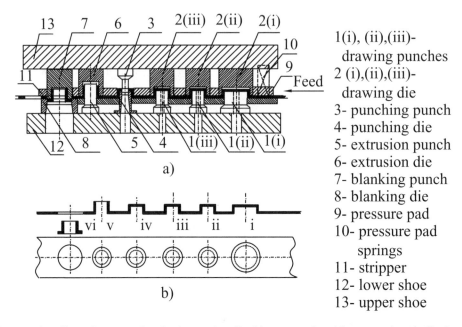

1(i), (ii),(iii)- drawing punches
2 (i),(ii),(iii)- drawing die
3- punching punch
4- punching die
5- extrusion punch
6- extrusion die
7- blanking punch
8- blanking die
9- pressure pad
10- pressure pad springs
11- stripper
12- lower shoe
13- upper shoe

Fig. 11.14 Progressive die and scrap strip: a) progressive die; b) scrap strip with successive shells drawn in this die.

A drawing, with successive reductions of the shell's diameter, is developed at the first three stations with punches 1(i), 1(ii), and 1(iii) attached to the lower die block set, and drawing rings 2(i), 2(ii), and 2(iii) attached to the upper shoe (13). At the fourth station, the bottom is punched with the punch (3), and punching die (4). In the fifth station the bushing wall is extruded to the final dimensions with the punch (5) and the extrusion die (6).

At the sixth station, the flange is blanked with the blanking punch (7) and blanking die (8). The final piece is dropped through the hole in the lower die block.

The pressure pad (9) functions as the upper stripper when the press slide is moved up. The lower stripper (11) allows all operations to be undertaken without interference between the die components and the workpiece. The die is equipped with a mechanism for advancing the strip stock (the mechanism is not shown in drawing).

Progressive dies for drawing by the second method — cut and carry — are of similar design to dies for drawing by the first method. The layout for drawing a similar bushing, where metal movement from the strip into the cup must be allowed for, is shown in Fig. 11.15.

An I-shaped relief cutout is notched at the first station of the die in Fig. 11.15. The cup is successively drawn at the second and the third stations. During all the drawing stages, metal is pulled in from the cutout edges as well as from the edges of the strip, thus narrowing the strip width. At the fourth station, the bottom hole is punched. At the fifth station, the bushing is extruded to its final dimensions. At the sixth station the flange is blanked, and the workpiece is separated from the strip and pulled out from the station. This type of relief cutout can be used when a series of shallow draws needs to be made without wrinkling the strip skeleton.

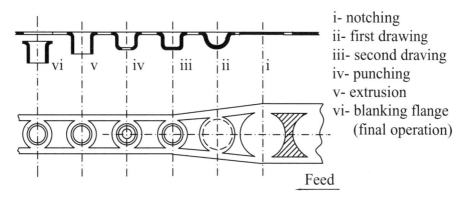

i- notching
ii- first drawing
iii- second draving
iv- punching
v- extrusion
vi- blanking flange
   (final operation)

Feed

Fig. 11.15 The scrap strip with successive shells drawn in a "cut and carry" die.

11.8 DRAWING DIES FOR SPHERICAL AND PARABOLIC SHAPE SHELLS

General problems in drawing these kinds of pieces include a tendency to wrinkles appearing at the top flange of the cup, and thinning of the material at the lower part of the drawn pieces. To avoid these problems, the following are necessary:

- The blank diameter needs to be a little larger than calculated.
- The excess the material must be cut off after each drawing operation.
- The order of drawing must be divided into two stages.

In the first stage, the workpiece is drawn with a spherical bottom, with or without a flange. In the second stage of the drawing operations, two methods are used:

- Reverse drawing
- Drawing in the die with a circular draw bead

Figure 11.16 shows a design for a die for drawing spherical pieces. In the first operation, a piece with a convex bottom is drawn. Then a reverse drawing method is used for the final drawing of the spherical piece. This method is used for drawing different shapes of pieces that must have a very fine surface without wrinkle marks. With this process, wrinkles do not form because the forces of compression encountered during the drawing operation are changed to forces of tension.

The technical components of the reverse die are as follows: the die redrawing ring (2), with a die ring radius which should be not less than 4 times the material thickness; a pressure pad (3); a reverse drawing punch (1); and a workpiece ejector (4). If the reverse method is used for drawing a parabolic piece, the final dimension of the workpiece sometimes needs to be developed by a spinning process.

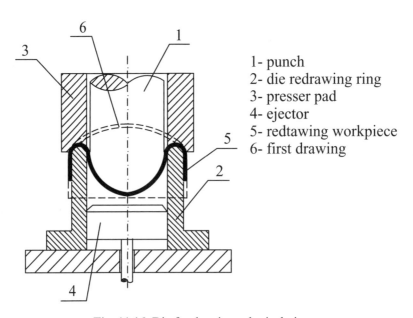

1- punch
2- die redrawing ring
3- presser pad
4- ejector
5- redtawing workpiece
6- first drawing

Fig. 11.16 Die for drawing spherical piece.

Figure 11.17 shows a design for a die with a circular draw bead ring. The components of the die are the die redrawing ring (2), the punch (1), the pressure pad (3), and the ejector (4). The die is used on a double-action press. The first blank forms a spherical shell in a separate die, which is then drawn in the die, as shown in Fig. 11.17. The punch is attached to the inner slide and the pressure pad to the outer slide. The pressure pad puts pressure on the spherical workpiece, so that the punch forms it into a parabolic shape. The workpiece is held with a force great enough to prevent the material from rising and light enough that the material is able to slide out from under the pad without wrinkles being formed.

The ejector (4) is designed with a parabolic inside contour matching the shape of the final piece so that it performs a coining die function before the workpiece is ejected from the die. This method is much more reliable than the reverse drawing method. The next radius reduction, which results in the final dimensions, is done by a spinning process.

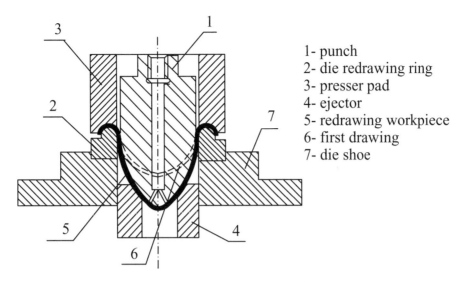

1- punch
2- die redrawing ring
3- presser pad
4- ejector
5- redrawing workpiece
6- first drawing
7- die shoe

Fig. 11.17 Die with a circular draw bead ring.

11.9 IRONING DIES

Frequently, the final operation in a series of draws consists in ironing the shell walls to reduce the thickness of the material and ensure a smooth uniform surface throughout. This work is done by making the clearance between the punch and the die ring slightly less than the thickness of the workpiece wall, so that the material is both thinned and elongated. Figure 11.18 shows an ironing die.

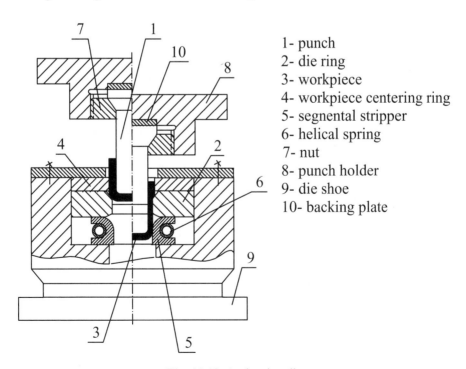

1- punch
2- die ring
3- workpiece
4- workpiece centering ring
5- segnental stripper
6- helical spring
7- nut
8- punch holder
9- die shoe
10- backing plate

Fig. 11.18 An ironing die.

The ironing die consists of the punch holder (8) and the die shoe (9), without a guide system. The punch is attached to the punch holder by the nut (7), whose inner side has a conic profile and whose outside is threaded. The upper set of the die is fixed to the ram by a clamp. In the die shoe (9) are fixed the die ring (2), the workpiece centering ring (4), and the segmental stripper (5), with the helical spring (6). The lower set of the die is attached to the bed of the press by the clamp. The segmental stripper consists of four segments connected by the ring of the helical spring (6), located in slots in the segments.

When the ram moves down, the segments are moved apart by the radial pressure on the workpiece, and when the ram is moved up, the sharp edges of the stripper strip the workpiece from the punch. The most efficient drawing process occurs when the first drawing achieves a reduction by diameter; reduction of the wall thickness of the workpiece occurs after that. Deformation by diameter undertaken separately from deformation by wall-thickness is advisable for two reasons. First is because of the favorable drawing ratio. Second is that for wall-thickness reduction operations, ironing dies are simple and may be used on single action presses. However, reduction by both diameter and wall thickness of the workpiece may be combined in one multi-stage drawing die. Figure 11.19 shows one design for a multi-stage ironing die.

In the multi-stage ironing die, a precut blank is inserted into a nest and held there by the pneumatic pressure pad (10). The blank is pushed through the die ring (2) for reduction by diameter, and then it is drawn. Ironing operations are then done by the punch (1) and three reduction draw rings (3, 4, and 5). The distance rings (6, 6a, and 6b) assure the correct distances between the draw rings. If in process of drawing, the workpiece leaves one draw ring before beginning to enter into the next draw ring, the machinery may perform another stroke that will often cause a crack or cracks in the workpiece.

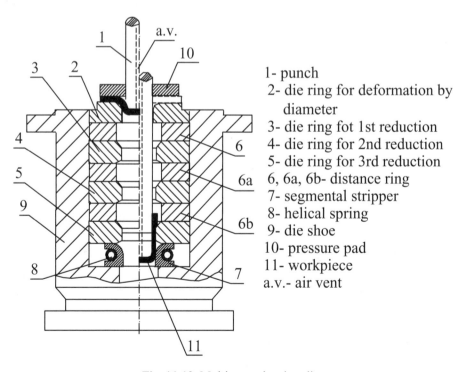

1- punch
2- die ring for deformation by
   diameter
3- die ring fot 1st reduction
4- die ring for 2nd reduction
5- die ring for 3rd reduction
6, 6a, 6b- distance ring
7- segmental stripper
8- helical spring
9- die shoe
10- pressure pad
11- workpiece
a.v.- air vent

Fig. 11.19 Multi-stage ironing die.

Twelve

Various Forming Dies

12.1 NOSING DIES

Deep drawn shells or tubes are used as initial working material in nosing type dies. Chapter 7 describes three types of nosing processes. Fig. 12.1 shows a design of a nosing die of Type I. The top of the workpiece diameter is reduced so that the workpiece, after nosing, has a conic shape. In this die, the nosing die ring (1) is fixed to the upper shoe so that it is movable, and the workpiece is held with an outside holder (2), attached to the lower shoe.

The workpiece is ejected by the ejector (3) or by knockout plate (4) after the nosing operation is finished.

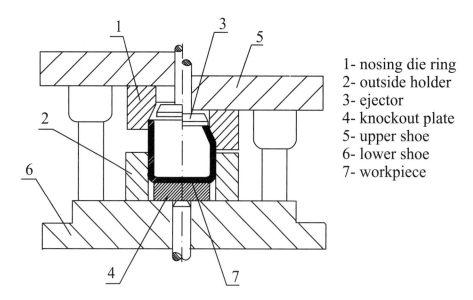

1- nosing die ring
2- outside holder
3- ejector
4- knockout plate
5- upper shoe
6- lower shoe
7- workpiece

Fig. 12.1 Schematic illustration of a Type I nosing die.

If the height of the workpiece is greater than the diameter, then deflections or wrinkles may appear in the workpiece during the nosing operation.

Figure12.2 shows a design for a Type II nosing die. To the upper shoe (9) is attached the nosing die ring (1), the inside support (6) with the spring (8), and the driver ring (5). To the lower shoe (10) are attached the outside support (3) and the segmental bushing (4) with helical springs (7). A pre-drawn workpiece is inserted into a nest on the outside support (3). When the press slide moves down, the driver ring (5) pushes the segmental ring (4), which then holds the workpiece. At the same time, the inside support (6) enters the workpiece to provide a positive guide and prevent deflection and development of wrinkles during the nosing operation.

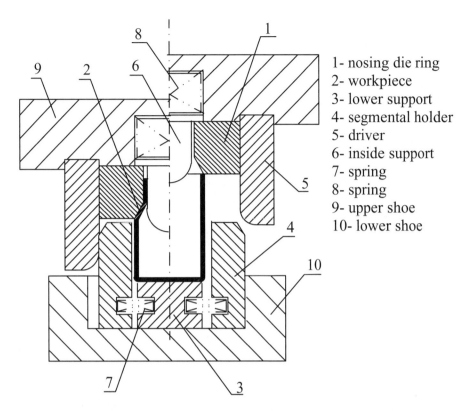

1- nosing die ring
2- workpiece
3- lower support
4- segmental holder
5- driver
6- inside support
7- spring
8- spring
9- upper shoe
10- lower shoe

Fig. 12.2 Schematic illustration of a Type II nosing die.

During the nosing operation, a die of Type III (as shown in Fig. 12.3) provides the best location for the workpiece.

The Type III die consists of the upper shoe (7) and the lower shoe (8). To the upper shoe is attached the nosing die ring (1), and to the lower shoe is fixed the outside support (3). The workpiece, which has been pre-drawn, is inserted into a nest on the outside support (3) and located by the segmental holder (4), which is powered by a mechanism positioned below the bed of the press. The workpiece is not then supported from the inside (Fig. 12.3a). Fig. 12.3b shows a similar die design for nosing a tube, but this type of die has an inside support (5) that holds the workpiece during the nosing operation.

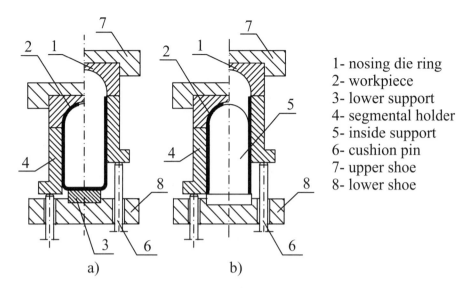

1- nosing die ring
2- workpiece
3- lower support
4- segmental holder
5- inside support
6- cushion pin
7- upper shoe
8- lower shoe

Fig.12.3 Schematic illustration of a Type III nosing die: a) without inside support of workpiece; b) with inside support of workpiece.

A complete nosing die design is shown in Fig. 12.4. The die consists of the upper shoe (8) and the lower shoe (9), guided by a guide post and guide post bushing. To the upper shoe is attached the nosing die ring (1), and the driver (4). To the lower shoe is fixed the inside support (5), the slide (6) with handle (7), and the segmental cam slide (3).

To place the workpiece on the inside support (5), the slide (6) needs to be pulled out of the work zone of the die before the nosing operation begins. The stop pin (not shown) positions the slide for nosing, which is done in three phases. In the first phase, the top of the workpiece is formed as a conic shape with a central angle of 40 degrees. In the second phase, the top of the workpiece is formed as a conic shape with a central angle of 75 degrees. In the third and last phase, the top of the workpiece is formed into a final hemispherical shape.

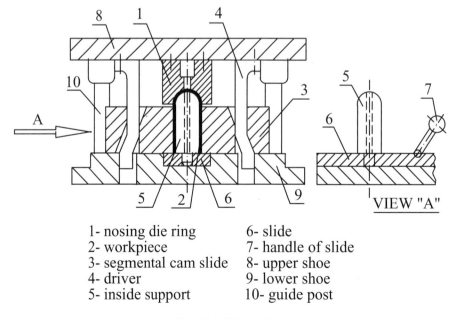

1- nosing die ring
2- workpiece
3- segmental cam slide
4- driver
5- inside support
6- slide
7- handle of slide
8- upper shoe
9- lower shoe
10- guide post

Fig. 12.4 Nosing die.

For all three phases, the same die is used except that the nosing die ring (1) is changed. When the press slide has gone down, the driver (4) pushes the segmental cam slides so they hold the outside of the workpiece. With the inside support (5), the die is provided with a positive location, resulting in a good quality workpiece. When the press slide is moved up, the driver pushes the segmental cam slides to the sides and the workpiece is freed.

12.2 EXPANDING AND BULGING DIES

Drawn shells of varying sizes and shapes, including tubular stock, can be expanded or bulged to produce such articles as teapots, water pitchers, kettles, doorknobs, parts of musical instruments, and various aircraft components.

12.2.1 Expanding Dies

Expanding dies are commonly used to enlarge the open end of a drawn shell or tubular stock with a punch. In most such operations, the workpiece is first annealed. Figure 12.5 shows a die for expanding one end of a tube (3). The die consists of a punch shoe (5) and a die shoe (6) with guide post (7). To the punch shoe is fixed the punch (1); the expanding die (8) and the knockout plate (2) are attached to the die shoe.

After the tube has been inserted in the die, the punch moves downward and expands the end of the tube. When the press slide is moved up, the knockout plate (2) pulls the workpiece (4) from the die (8).

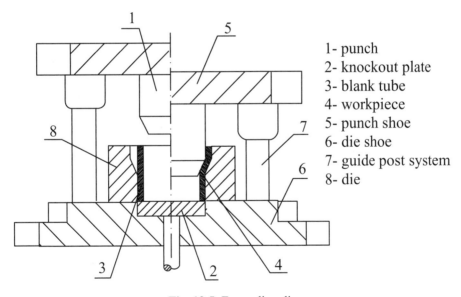

1- punch
2- knockout plate
3- blank tube
4- workpiece
5- punch shoe
6- die shoe
7- guide post system
8- die

Fig. 12.5 Expanding die.

12.2.2 Bulging Dies

Vertical or horizontal segmented dies are commonly used for bulging, the force being applied by either hydraulic or mechanical means. Figure 12.6 shows a schematic design for a mechanical bulging die with a segmental punch for bulging a ring of a tube. This die does not have a die cavity, the tube being pulled on the segmental punch (1). The force is applied to the segmental punch (1) by the ram pressing down onto the support plate (3).

The segmental punch slides down over the cone (2) and forms the tube into its final shape. When the press slide moves up, the stripper plate lifts the segmental punch, which is pulled inward by springs, so that the workpiece is freed. This type of bulging die is used for the production of symmetrical cylindrical components.

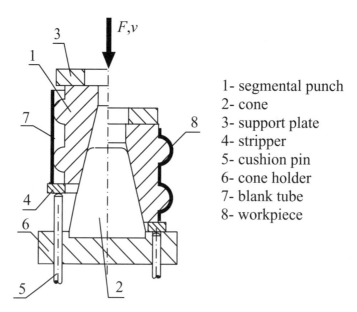

1- segmental punch
2- cone
3- support plate
4- stripper
5- cushion pin
6- cone holder
7- blank tube
8- workpiece

Fig. 12.6 Mechanical bulging die.

Figure 12.7 shows a schematic design for a bulging die with an elastic insert. The form of the finished piece is machined into a split die (1) so that, when pressure is applied to the rubber insert (2), the workpiece is forced into the forming cavity.

When the press slide is moved up, the die is opened and the rubber insert is removed from the component. For this type of die, medium-hard rubber or polyurethane is used and is easy to handle. The major advantage of using polyurethane inserts is that the inserts are resistant to abrasion, water, and lubricants. Furthermore, they do not damage the surface finish of the piece.

Hydraulic means can also be used for bulging operations, but they require sealing and control of the hydraulic pressure.

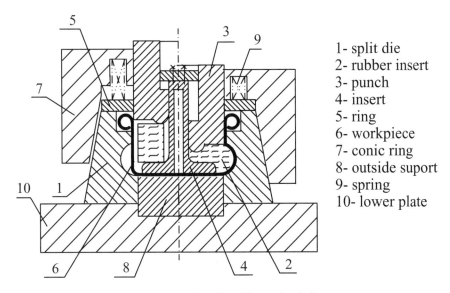

Fig. 12.7 Bulging die with an elastic insert.

1- split die
2- rubber insert
3- punch
4- insert
5- ring
6- workpiece
7- conic ring
8- outside suport
9- spring
10- lower plate

12.3 FLANGING DIES

There are two kinds of flanges: the convex-shrink flange and the concave-stretch flange. The convex flange is subjected to compressive hoop stresses, which, if excessive, cause the flange edges to wrinkle. In concave flanging, the flanges are subjected to tensile stresses which, if excessive, cause cracks at the edges.

This section discusses only dies for the flanging of a hole; they can be single-operation dies or combination dies. Figure 12.8 shows a single-operation die for flanging a hole on the bottom of a drawn shell.

The die consists of the upper shoe (7) and the die (lower) shoe (8), which are located by the guide post (3). The flanging die ring (1) and the ejector (6) are attached to the die shoe. The flanging punch (2) and the pressure pad (4), with the springs (5), are fixed to the punch holder. When the angle of the flange is less than 90 degrees, the process is called dimpling. The dimpling operation has been used very extensively in aircraft production.

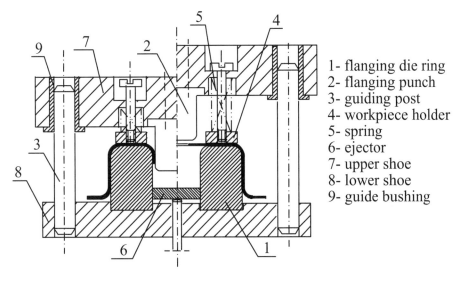

1- flanging die ring
2- flanging punch
3- guiding post
4- workpiece holder
5- spring
6- ejector
7- upper shoe
8- lower shoe
9- guide bushing

Fig. 12.8 Flanging die.

A combination die for punching, blanking, and flanging is shown in Fig. 12.9. The die consists of the punch holder (12) and the die shoe (13); they are guided by posts and bushings (not shown).

The punch holder carries the punch (1), a blanking punch (2), a pressure pad (9), and knockout pins (11), plate (10), and knockout pin plate (3).

To the die shoe are attached the blanking plate (4), the punching ring (6), the workpiece ejector plate (5), and the strip stop pin (8).

The strip stock is positioned by the stop pin (8), and guided by a guide rail (not shown) and held by the pressure pad (9), which functions as a scrap stripper. First, the blank is cut out with the punch (2) and the blanking plate (4), so that the hole is simultaneously punched with the punch (1), and the punching die ring (6). The hole is then flanged with the inside of the blanking punch (2) and the outside of the punching die ring (6). The workpiece is ejected from the die by the knockout plate (3) and the workpiece ejector (5).

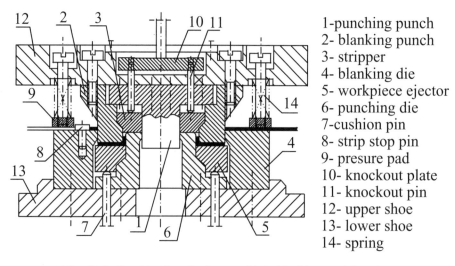

1-punching punch
2- blanking punch
3- stripper
4- blanking die
5- workpiece ejector
6- punching die
7-cushion pin
8- strip stop pin
9- presure pad
10- knockout plate
11- knockout pin
12- upper shoe
13- lower shoe
14- spring

Fig. 12.9 Combination die for punching, blanking, and flanging.

Thirteen

Tool and Die Materials

13.1 INTRODUCTION

The composition and physical properties of the principal materials used in die design for sheet metal form-ing are discussed in this chapter. A die component must be made of a material that has properties suitable for the conditions of service. The die material is just as important as are considerations of the loads and stresses, dimension, and form of the piece, as well the quantity of the pieces which need to be manufactured with the die. Frequently, the limitations imposed by available materials are the controlling factor in a die design. The designer must be familiar with the effects that the methods of manufacture of die components and heat treatment have on the properties of the die materials. The manufacturing processes used for fabrication of workpieces will also influence the types of material that can be used. Sometimes, the help of a professional metallurgist is needed to ensure the best possible choice of material and heat treatment. The emphasis of this text will be on the steels needed to produce dies for sheet-metal forming. Other materials, such as nonferrous and nonmetallic materials, will be included to complete the picture.

13.2 CARBON AND ALLOY STEELS

Steel is an alloy of iron and carbon, or of iron, carbon, and other alloying elements. Carbon must be present to the extent of about 0.05% by weight in order for the material to be known as steel rather than commercial iron.

Carbon and alloy steels are among the most commonly used metals and have a wide variety of applications. The composition and processing of steels are controlled in a manner that makes them suitable for numerous applications. They are available in various basic product shapes: plate, sheet, strip, bar, wire, tube, and they may also be cast or forged.

13.2.1 Designations for Carbon and Alloy Steels

The numbering system of the Society of Automotive Engineers (SAE) and the American Iron and Steel Institute (AISI) is based on the chemical composition of materials and provides a simple means whereby any particular steel can be specified. In general, this catalog system uses a number composed of four digits. The first two digits give the type or alloy classification; the last two (and in some instances, three) digits give the carbon content. For example, steel 1020 indicates plain carbon steel containing 0.20% carbon.

The basic designations by AISI and SAE for carbon and alloy steels are given in Table 13.1

Table 13.1 Basic designation for SAE and AISI steels

Carbon steels:	lxxx	Chromium steels:	5xxx
- Plain carbon steel	10xx	- Low chromium	51xx
- Resulphurized carbon steel	11xx	- Medium chromium	52xx
Manganese steels	l3xx	Chromium-vanadium steels: 1.00%Cr	6xxx 6lxx
Nickel steels: 3.50%Ni 5.00%Ni	2xxx 23xx 25xx		81xx 86xx
Nickel-chromium steels: 1.25% Ni, 0.60% Cr 1.75% Ni, 1.00% Cr 350% Ni, 1.50% Cr	3xxx 3lxx 32xx 33xx	Chromium-nickel-molybdenum steels:	87xx 88xx
Molybdenum steels: - Carbon-molybdenum steels - Chromium-molybdenum - Chromium-nickel-molybdenum - Nickel-molybdenum 1.75% Ni - Nickel-molybdenum 3.50% Ni	4xxx 40xx 4lxx 43xx 46xx 48xx	Silicon-manganese steels 2.00% Si	9xxx 92xx

13.2.2 Effects of Various Alloy Elements in Steels

Various elements are added to steels to impart properties of hardenability, strength, wear resistance, workability, and machinability. In Table 13.2, these elements are presented with both their beneficial and detrimental effects.

In Table 13.2, the sign "+" connotes positive effects, and "−" means negative effects. Double marks mean greater effects.

13.2.3 Carbon Steels

Carbon steels are generally classified in the fallowing ways:

Law-carbon steel. Also called mild steel, low-carbon steel has less than 0.30% carbon. Structural flat steel — both cold-rolled and hot-rolled — is used in the construction of large or small jigs and fixtures. These steels need not have high strength or wear-resistant properties. They may be carburized and case-hardened if better mechanical properties are desired. Cold-rolled steels have a good finish and therefore need not be machined, whereas hot-rolled steel has an oxide scale surface and must be machined if a smooth surface is required. It should be noted that rolling stresses are locked into the surface of cold-rolled steel.

Table 13.2 Influence of alloying elements on the characteristic properties of steels

Properties of steels	Alloying elements in steels											
	C	**S**	**P**	**Si**	**Mn**	**Ni**	**Cr**	**Mo**	**W**	**V**	**Ti**	**Co**
Strength	++		+	+	+	+	+	+	+	+		+
Hardness	++			+	+	+	++	+	+	+		+
Tensile	−	−		++	−−	+	+					−
Elasticity	++		+	++	+	+	+			++		
Impact strength	−	−−		−	++	+		++		++	−−	
Hardness and strength at elevated temp.												
Dynamics strength												
Corrosion resistance	−	−−	++	+	+	++	+++	+		+	+	+
Formability	−−	−	−−	−−	−	−−	−		−	−−		
Machinability	−	++	+	−	−	−	−		−	−	−	
Wear resistance	+				+		+	+		+		
Weldability	−	−	−	−	−							+
Toughness			−			+	+	+				

Medium-carbon steel. Medium-carbon steel has a carbon content of between 0.30% and 0.70%. It is generally used in applications requiring higher strength than low-carbon steel, such as automotive, machinery, and railroad equipment, and parts such as gears and axles.

High-carbon steel. High-carbon steel has more than 0.70% carbon. It is generally used for parts requiring high strength and high hardness. The higher the carbon content of the steel, the higher its hardness, strength, and wear resistance after heat treatment.

13.2.4 Alloy Steels

These kinds of steels contain significant amounts of alloying elements and are usually made with more care than are carbon steels. Alloy steels are used in applications where strength, hardness, creep/fatigue resistance, and toughness are required. These steels may also be heat-treated to obtain the desired properties.

13.2.5 Machinability of Steels

The relative ease with which a given material may be machined, or cut with sharp-edged tools, is called machinability. Machinability ratings are based on a tool life of T = 60 min. The standard is AISI 1112 steel, which is given a rating of 100. Thus, for a tool life of 60 minutes, this steel should be machined at a cutting speed of 100 ft/min (0.5m/s). A higher speed will reduce tool life, and lower speeds will increase it. For example, tool steel AISI A2 has a machinability rating of 65. This means that when this steel is machined at a cutting speed of 65 ft/min (0.325 m/s), tool life will be 60 minutes. Some materials have a machinability rating of more than 100. Nickel has a rating of 200, free-cutting brass of 300. In Table 13.3, ratings are given for some kinds of carbon and alloy steels.

Table 13.3 Machinability rating for some types of steels

AISI	1010	1020	1040	1060	3140	4340	6150	8620
Rating	55	65	60	53	55	45	50	60

13.2.6 Mechanical Properties of Steels

Typical mechanical properties of selected carbon and alloy steels are given in Table 13.4.

Table 13.4 Mechanical properties of selected carbon and alloy steels

AISI	CONDITION	UTS (MPa)	YS (MPa)	Elongation (%)	Hardness (HB)
1020	As-rolled	448	330	36	143
	Normalized	441	346	35	131
	Annealed	393	294	36	111
1040	As-rolled Normalized	620	413	25	201
	Annealed	589	374	28	170
		518	356	30	149

(Continued)

AISI	Condition	UTS	YS	El	HB
1060	As-rolled Normalized	813	482	17	341
	Annealed	775	420	18	229
		625	372	22	179
1080	As-rolled Normalized	965	586	12	293
	Annealed	1010	524	11	293
		615	375	24	174
3140	Normalized	891	559	19	262
	Annealed	689	422	24	197
4340	Normalized	1279	861	12	229
	Annealed	744	472	22	197
6150	Normalized	939	615	21	269
	Annealed	667	412	23	197
8620	Normalized	632	357	26	189
	Annealed	536	385	31	149
8620	Normalized	632	357	26	189
	Annealed	536	385	31	149

Table 13.5 gives typical mechanical properties of selected carbon and alloy steels in quenched and tempered condition.

Table 13.5 Mechanical properties of quenched and tempered carbon and alloy steels

AISI	Tempering temperature (°C)	UTS (MPa)	YS (MPa)	Elongation (%)	Hardness (HB)
1040	205	779	593	19	262
	425	758	552	21	241
	625	634	434	29	192
1060	205	1103	779	13	321
	425	1076	765	14	311
	650	800	524	23	229
1080	205	1310	979	12	388
	425	1289	951	13	375
	650	889	600	21	255
4340	205	1875	1675	10	520
	425	1469	1365	10	430
	650	965	855	19	280
6150	205	1931	1689	8	538
	425	1434	1331	10	420
	650	945	841	17	282

13.2.7 Applications of Carbon and Alloy Steels

Characteristic and typical applications of various carbon and alloy steels are given in Table 13.6.

Table 13.6 Typical applications and characteristics of various carbon and alloy steels

TYPE	AISI	Condition	Typical applications
Low-carbon (Carburizing Grades)	1117 1020 1030 4320 8620 9310	Soft	Poor abrasion resistance and metal-to-metal wear resistance do not hold a cutting edge or sustain high loads. They provide satisfactory service life as pins, guides, shafts, etc.
		Carburized to>600 HB	Excellent metal-to-metal wear resistance. Suitable for guide posts and guide rails
Medium-carbon (Direct Hardening)		Soft	Same as low carbon. Widely used for parts requiring good strength and toughness:
High-carbon (Direct hardening)	52100 1080 1095	Soft	The low stress abrasion and metal-to-metal wear resistance are better than with soft low- and medium-carbon steels. 1080 strip at 450 HB can be used for Rule dies.
		Hardened and tempered to>500 HB	52100 at 655 – 680 HB is standard steel for rolling elements. It also has suitable wear properties for short-run dies. 1080 and 1095 steels in strip are widely used for flat springs.

13.3 TOOL AND DIE STEELS

Tool and die steels are specially alloyed steels that are designed for high strength, impact toughness, and wear resistance. They are commonly used in the forming and machining of metals at both room and elevated temperatures.

The steel for most types of tool and dies must be in a heat-treated state, generally hardened and tempered, to provide the properties needed for the particular application. Thus, tool and die steels must be able to withstand heat treatment with a minimum of harmful effects, dependably resulting in the intended beneficial changes in material properties.

13.3.1 Designation and Classification of Tool and Die Steels

The designation and classification system established by AISI and SAE for tool and die steels has seven basic categories. These categories are associated with the predominant application characteristics of the tool and die steel types they comprise.

A few of these categories are composed of several groups to distinguish between families of steel types that, while serving the same general purpose, differ with regard to one or more dominant characteristics. Table 13.7 summarizes the basic types of tool and die steels.

Table 13.7 Basic types of tool and die steels

TYPE	AISI	BASE CATEGORIES
High speed	M	Molybdenum base
	T	Tungsten base
	H1 to H19	Chromium base
Hot work	H20 to H39	Tungsten base
	H40 to H59	Molybdenum base
Cold work	D	High carbon, high chromium
	A	Medium alloy, air hardening
	O	Oil hardening
Mold steels	P1 to P19	Low carbon
	P20 to P39	Others
Mold steels	L	Low alloy
	F	Carbon - tungsten
Water hardening	W	

13.3.2 Cold Work Tool and Die Steels

Cold work steels (A, D, and O steels) are used for cold working operations as well as for sheet-metal forming operations, and they are described in more detail in this section. They generally have high resistance to wear and cracking. These steels are available as oil-hardening or air-hardening types.

Major alloying elements. Typical analysis results of alloying elements for these types of steels are given in Table 13.8.

Table 13.8 Typical alloying elements for cold work tool and die steels

AISI	Alloying elements by percentage								
	C	Mn	Cr	Mo	V	Si	Ni	W	Co
A2	1.00	0.60	5.00	1.00	0.30	0.25			
A4	1.00	2.00	1.00	1.00		0.25			
A8	0.55	0.25	5.00	1.50	0.25	1.00		1.25	
A.9	0.50	040	5.00	1.50	1.00	1.00	1.50		
D2	1.50	0.30	12.00	0.75	0.60	0.25			
D3	2.25	0.30	12.00	0.80	0.60	0.25			
D7	2.35	0.40	12.50	1.00	4.00	0.40			
O1	0.90	1.25	0.50			0.25	0.25	0.50	
O2	0.90	1.60	0.25	0.30	0.20	0.25			
O6	1.45	0.25	0.50	0.25	0.25	0.25		1.55	

Characteristics and applications. Machinability and typical applications of common die steels are given in Table 13.9.

Table 13.9 Typical applications of common tool and die steels

AISI	Machinability Rating	Approximate Hardness (HRc)	Characteristics and Applications
A2	60	57 to 62	Good combination of wear resistance and toughness, good size stability in heat treatment, good hardenability. Typical applications involve blanking and forming dies, punches, and forming rolls.
A6	65	57 to 60	Air hardening from low temperature, excellent size stability in heat treatment, deep hardening. Typical applications involve blanking and forming dies, punches, coining and bending dies and plastics molds.
A8	70	56 to 59	Optimum combinations of wear resistance and toughness, superior size stability, suitable for highly abrasive hot work requirements. Applications include shear blades, trim dies, forging dies, and plastics molds.
A9	65	35 to 56	Medium resistance to decarburization and to wear. Solid cold heading dies, die inserts, coining des, forming dies, punches, and rolls.
D2	50	57 to 62	Very high wear resistance. Excellent size stability, deep hardening in air. Typical applications include blanking dies, drawing dies, shear blades, forming rolls, and trim dies.
D3	35	57 to 64	Excellent abrasion resistance, very high compressive strength, high hardening response. Typical applications include blanking dies, drawing dies, shear blades, forming rolls, punches, and cold trimming dies
D5	50	59 to 63	Very high wear resistance, optimum size stability, and superior resistance to tempering. Applications include blanking dies, shear blades, forming rolls, hot and cold punches, swaging dies, and cold trimming dies.
O1	90	59 to 61	Moderate wear resistance, relatively safe to harden, easy to machine. Common applications include blanking and forming dies, bending and drawing dies, plastics molds, and shear blades.
O2	fairly	57 to 61	Fairly good wear resistance, relatively safe to harden, fairly easy to machine. General-purpose tooling with a combination of wear resistance and moderate toughness.
O6	125	45 to 63	Excellent machinability, outstanding resistance to wear and galling, easy hardening. Typical applications are gages, punches, cold forming, blanking, and trimming dies, bushings, and other machine tool parts.

13.4 NONFERROUS METALS

Many nonferrous metals are used in tool and die, jig, and fixture design. Such materials include alloys of aluminum, magnesium, brass, bronze, zinc-base alloys, and beryllium. Although more expensive than ferrous metals, nonferrous metals and alloys have important applications because of their numerous positive characteristics, such as low density, corrosion resistance, ease of fabrication, and color choices.

Aluminum alloys. The most important factors in selecting aluminum alloys for use in tool and die design are their high strength-to-weight ratio, ease of machinability, resistance to corrosion by many chemicals, and non-magnetic properties.

Aluminum alloys of series 2024 and 7075 are two of the more widely used alloys for temporary dies, limited production runs, fixture bodies, or other special purposes.

Copper alloys. Copper alloys are often attractive for applications where a combination of desirable properties, such as strength, corrosion resistance, thermal conductivity, wear resistance, and lack of magnetic polarization are required. The most common copper alloys used in tool and die design are bronzes. Bronze is an alloy of copper and tin. There are also other bronzes, such as aluminum bronze, which is an alloy of copper and aluminum; beryllium bronze; and phosphor bronze. Aluminum bronze is most widely used in die design for guide bushings.

Zinc-based alloys. These alloys are used extensively in die-casting. They may be cast into shapes quickly for the purposes of being used as short-run punches and dies for either short-run production or for experimental short runs.

One of the better-known zinc-based die materials is Kirksite. Kirksite is a nonferrous alloy for the production of press tooling, sheet metal forming, and plastics molds. Kirksite is a zinc-based alloy of 99.99 % pure zinc and contains precise amounts of alloying elements that give it very high impact strength. It is also very free flowing when molten. Kirksite has good machinability and abrasion resistance with relative freedom from loading and galling (wearing away by friction), and it can quickly be polished to a high surface finish. It is very useful for blanking dies using no reinforcement at all, the Kirksite being simply cast around a steel punch that may have a very complex profile. Another advantage of this material is that it may be remelted and used over again.

Bismuth alloys. These low-melting alloys expand upon solidification, especially those with high bismuth content. This characteristic of bismuth makes it possible to use these alloys for duplicating mold configurations that would otherwise require many operations to reproduce. The material is suitable for making forms used in stretch forming and has also been used for other components in punches and dies.

13.5 NON-METALLIC MATERIALS

In addition to the ferrous and nonferrous metals, there is a wide variety of non-metallic materials that are important to the tool designer. Several of these are discussed in this text.

Plastics. Plastics are composed of polymer molecules and various additives. Polymers are long-chain molecules formed by polymerization, that is, by linking and cross-linking of different monomers. Thermosetting plastics such as epoxy, polyester, and urethane are used widely as tooling material. These thermosets cure (harden) at room temperature. Although curing takes place at ambient temperatures, the heat of the reaction cures the plastics. To impart certain specific properties for tooling material, polymers are usually compounded with additives. These additives modify and improve certain characteristic of plastics, such as their stiffness, strength, hardness, abrasion resistance, dimensional stability, color, etc. In some instances, steel wear plates are inserted in the plastics. Plastics tooling is used in many operations, such as: drawing and forming dies, drill jigs, assembly tools, and inspection fixtures. These materials have advantages over many other materials. They are resistant to chemicals, moisture, and temperature. Generally they are easy to machine, rework, and modify.

Elastomers comprise a large family of amorphous polymers having a low glass-transition temperature. They have a characteristic ability to undergo large elastic deformations without rupture; they are also soft and have a low elastic modulus.

The terms "rubber' and "elastomer" are often used interchangeably. Generally, an elastomer is defined as being capable of recovering substantially in shape and size after a distorting load has been removed.

Rubber is defined as being capable of recovering from large deformations quickly. A rubber pad confined in a container is used as the forming die. Operations such as forming, blanking, bulging, and drawing may be done with a rubber or polyurethane member as one of the components of the die set.

Polyurethane has very good overall properties of high strength, stiffness, hardness, and exceptional resistance to abrasion, cutting, and tearing. Typical applications as a tool material include cushioning and diaphragms for rubber forming of sheet metals.

Fourteen

Quick Die-Change Systems and Die Design

14.1 INTRODUCTION

The sheet metal stamping industry used to enjoy long production runs, high inventory levels, and extended die change times. However, with rapidly increasing changes in workpiece type and design, and smaller batch sizes, reducing the die setup time is of crucial importance for the profitability of many companies.

In the early 1990s, many businesses in the United States and Western Europe began looking seriously at Just-in-Time, a Japanese management philosophy that seeks to improve profitability by eliminating the delays inherent in long production runs and high inventory levels. Although Just-in-Time can deliver many benefits, it demands smaller production quantities to meet customer schedules. Those smaller runs mean more set-ups are required per day or week than were necessary even ten years ago. More set-ups usually mean more non-productive time and less time during which a press is being utilized. Modern management views non-productive time as lost profits; they must seek to minimize this lost time if they wish to remain competitive.

In the sheet metal stamping environment, one place where time can profitably be cut is in the change-out of a worn-out die for a new one. The die exchange can be defined as all of the work necessary to change from the last good part on one die to the first good part on the next one. The time for this change is measured starting from the moment the last good part from the first run is completed until the first acceptable part of the second run is produced. Just-in-Time management has brought in the idea of quickly changing out dies to save time.

A logical component of Just-in-Time is Quick Die Change (QDC), whose goal is to be able to change dies out in a single minute (single minute exchange of dies, known as SMED).There are those who advocate SMED. However, this goal is not attainable in practical terms today, though perhaps at some time in the future it may be. At present, there are stamping shops that are comfortable with die change times of one to two hours or more. To them, SMED is unimaginable, and even a 50-percent time reduction is excellent. But there are some shops that are achieving a 20-minute or less die change today and hope to extend the goal to ten minutes or even less.

14.1.1 Quick Die Change Benefits

Productivity. The most obvious benefit of QDC is an increase in productivity. This productivity gain is accomplished by reducing the downtime associated with changeovers. In some cases, the number of presses needed to produce a part is reduced, and operator overtime can be eliminated.

Reduction of inventories. Quick die changes allow a metal-forming shop to run smaller quantity runs and change dies more frequently to meet actual demand. When die changes require too much time, a plant operator can't afford repeated downtimes and must instead make longer runs for anticipated demand, increasing storage costs for inventory of unsold products.

Longer tooling life. When dies are handled in a controlled and safe manner, there is less chance for damage. When dies are placed consistently in the same press location, the quality of the parts also becomes more consistent.

Increased capacity. With less downtime for die exchanges between the last part and the first part, presses can meet the production demands of several customers quickly without resorting to overtime or added shifts, thus increasing revenue.

Improved Quality. The automation that a quick die-change system requires means it must accurately place a die in the same position each time it is placed back in the press. That accuracy in repeatable positioning means less waste.

Lower labor cost. A quick die-change system guides the die and rolls it smoothly in and out of presses during the exchange. Often only a single operator controls the process, and the system won't allow a press to operate if a die is not placed in its proper position.

Improved safety. Employee safety also can be improved with consistent quick die change procedures and equipment designed specifically for the task. The improved work environment can lead to better attitudes in general—an intangible factor, but an important one.

14.2 QUICK DIE-CHANGE SYSTEM CONFIGURATION

In addition to the study of sheet metal forming processes and die design, it is helpful to have an understanding of the construction and configuration of the various quick die-changing systems. There are many types of die-changing equipment, but all QDC components systems fall into three main categories:

1. die clamping systems
2. die transform and positioning systems
3. die transport systems

14.2.1 Die Clamping Systems

QDC clamp components can be mechanical, hydraulic, or permanent electro-mechanical clamps.

QDC clamps can be secured quickly to the press bolster and ram. Both clamping and unclamping can be accomplished automatically or with a small amount of manual effort. Although mechanical varieties are available, hydraulic clamps generally offer a higher degree of safety. Typically, if a hydraulic clamp is not positioned properly, the controls will receive a low pressure reading or electrical signal, and the press will not be allowed to operate. Hydraulic clamp components may be either single-acting or double-acting.

a) Single-Acting Clamps

There are many types of single-acting clamps but the most common types include the following:
- hydraulic rocker clamps
- hydraulic ledge clamps
- hydraulic T-slot sliding clams
- mechanical clamps

Hydraulic rocker clamp. A standard single-acting hydraulic rocker clamp is for mounting into a T-slot opening of a press or bolster plate. Figure 14 schematic illustration shows a single-acting hydraulic rocker clamp. This type of clamp employs a clamping force up to 250 kN, and stroke of 5 to 10 mm.

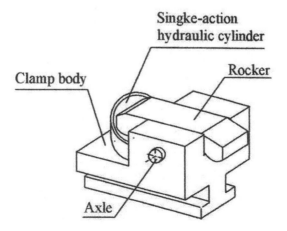

Fig. 14.1 Single-acting hydraulic rocker clamp.

Hydraulic ledge clamp. The hydraulic ledge clamp includes four or more modular single-acting spring return cylinders. The clamps are mounted at the side of the press table, which enables the die to be loaded from the operating side towards a fixed stop. With different support blocks, heights can be adapted to the die's clamping height. If the upper shoe is clamped at the ram of the press, the die half is prevented from falling. The capacity of this type of clamp ranges up to 120 kN, and a maximum of an 8-mm stroke of the cylinders. Figure 14.2 shows a single acting hydraulic ledge clamp with four cylinders.

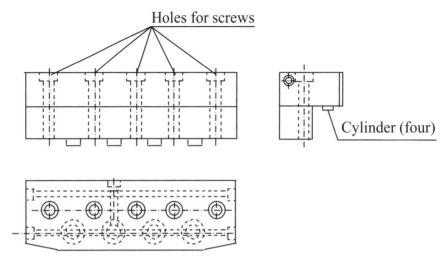

Fig. 14.2 Single-acting hydraulic ledge clamp.

Hydraulic T-slot sliding clamp. Hydraulic T-slot sliding clamps are single-acting manually positioned hollow cylinder-type clamps that are widely used on many QDC applications to clamp the lower half of a die to a bolster or press table, and the upper half of the die to the ram of the press. Some cylinders have a swivel plate, which allows their application to non-straight die surfaces.

Cylinder-type clamps have a greater capacity within their small dimensions due to their higher operating pressure of 350 MPa, and with an 8-mm stroke, they are suitable for many applications. Figure 14.3 schematically illustrates a single-acting hydraulic T-slot sliding clamp.

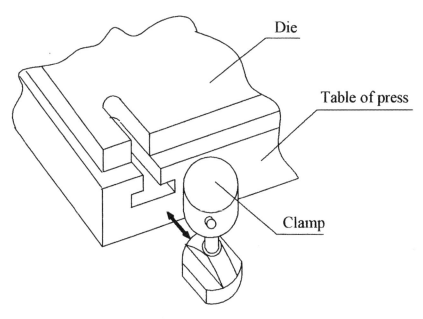

Fig. 14.3 Single acting hydraulic T-slot sliding clamp.

Mechanical clamp. The main characteristics of this clamp are the following: it is a safe, quick, and simple way to clamp a die; it features easy insertion into existing T-slots; it also has easy positioning without additional fixing; it achieves full pressure quickly and easily by turning the low torque "power nut" 180°; and it can self-lock into a clamped position. Figure 14.4 shows a mechanical clamp.

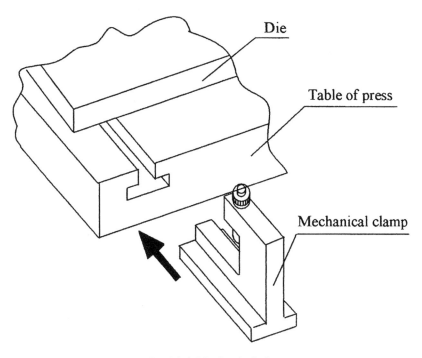

Fig. 14.4 Mechanical clamp.

b) Double-Acting Clamps

Because of the range of clamping forces (15 to 1200kN), double-action clamps can be applied to a large range of applications. Usually they have two sensors to detect the clamped and unclamped position of the plunger. The most common types of double-acting hydraulic clamps are the following:

- self-locking clamps
- swing clamps
- rod swing clams

Self-locking wedge clamps. Self-locking double-acting wedge clamps are the most commonly employed clamps for fully automated QDC systems. A typical clamping system consists of four double-acting cylinders mounted on the lower press bed. Due to the 5-degree wedge angle, these cylinders are self-locking even without hydraulic pressure, which is an important safety feature. For additional safety in QDC applications, continuous hydraulic pressure is used. Two safety sensors indicate the clamped and retracted position of the plunger. Figure 14.5 shows a double-acting self-locking wedge clamp.

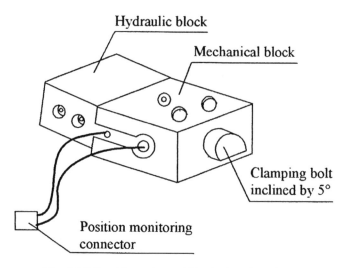

Fig. 14.5 Double-acting self-locked wedge clamp.

Swing clamp cylinder. The swing clamp's arm swings 90 degrees and then clamps vertically. These clamps are ideal for mounting beneath the surface bed or bolster surface to allow unobstructed access to a part or die loading and unloading. For the clamping arm, it is necessary for the dies to have T-slot openings. If the swing clamp is designed to have three positions on the arm piston, the clamps are integrated on the press table and on the moving bolster plate. Due to the fact that the clamping arm is retracted completely and disappears under the working table, the die and/or unloading surfaces become totally free.

Three-position swing clamps are particularly recommended when the application requires an integrated clamping solution for press tables or when there is a narrow space and limit access.

Piston rotation is obtained through a mechanical device located in the lower part of the clamp and activated by two independent hydraulic lines whereas the straight action of the piston is provided by two other hydraulic lines. The clamping arm's position (clamped/unclamped) is controlled by two proximity switchers whereas the straight stroke is controlled by pressure switches. Figure 14.6 shows a double-action three-position swing clamp.

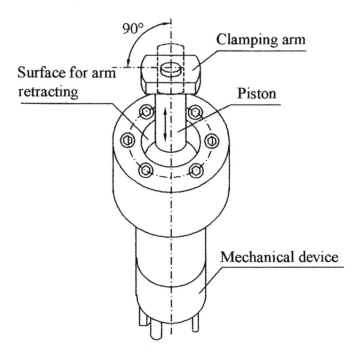

Fig. 14.6 Double-action three-position swing clamp.

Rod swing clams. Figure 14.7 shows a double action hydraulic rod swing clamp. These rod swing clamps can be used to clamp the upper die part. To have good access to the die, the rod swings out 45 to 90 degrees depending on the model. Two proximity switches indicate the clamped and unclamped positions of the clamp rod.

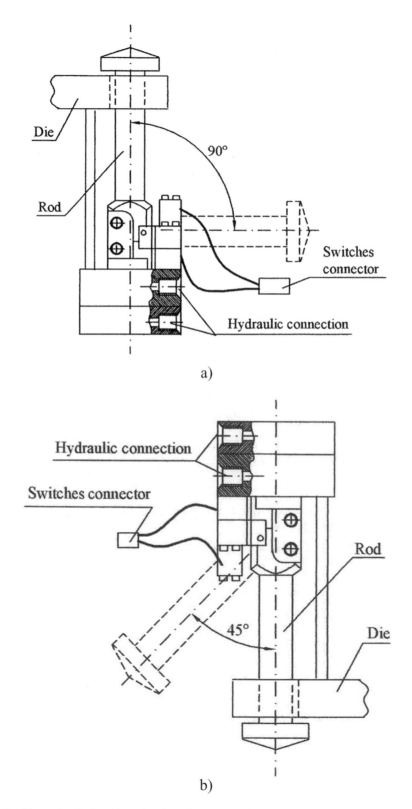

Fig. 14.7 Double-acting hydraulic rod swing clamp: a) rod swing 90 degree; b) rod swing 45 degree.

c) Permanent Electric Magnetic Clamps

A permanent electric magnetic clamp, used and known as a *pressmag* system, is based on 50 × 50 mm square invertible AlNiCo magnets surrounded by an isolated coil. On top is a round pole and each square is surrounded by non-invertible permanent rare earth magnets. The coil around the invertible magnets generates an electric magnetic field, which inverts the magnets within fractions of a second. Clamping of the die is finished for an unlimited time and without electrical energy or generating heat. Another electric pulse will demagnetize the system, releasing the die. After the demagnetization, the poles are neutral. Total holding force is directly proportional to the number of magnetic poles engaged with the die and taking into account the aforementioned specifications and work conditions. The primary advantage of magnetic clamping is that it avoids the need for standardized back plates. The system is mounted with bolts to existing T-slots, clamping claws or tapped holes. One operator can operate the die clamping in a safe way standing outside the press. Figure 14.8 shows a pressmag clamp.

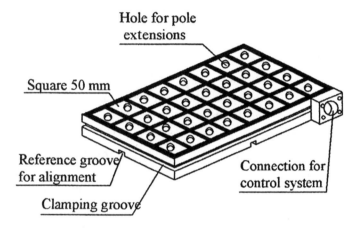

Fig. 14.8 Permanent electric magnetic clamp.

14.2.2 Die Transfer, Transport and Positioning Systems

This equipment grouping allows dies to be loaded into a press (or unloaded from a press) and moved into position quickly and accurately. It helps eliminate the dangers associated with traditional manual methods of die exchange and also reduces the risk of damaging dies during the handling process.

The transport equipment helps ensure the safe and timely movement of dies between storage locations and press side loading (and unloading) positions.

The transfer systems have the most effect on increasing the operational uptime of the sheet metal stamping processes. Many transfer systems are used, but the *rigid chain rolling bolster* and *die cart* systems are prime examples of such QDC equipment.

a) Rigid Chain Rolling Bolster Systems

These transform and positioning systems can be used for presses with bed heights up to 750 mm. The metal-forming industry uses many types of rolling bolster systems, but two are the most common, especially in smaller metal-forming shops. These are single- and double-bolster operation systems. Each system has hydraulic die lifters or hydraulic cylinders that are mounted to lift four bolster wills into functioning positions upon activation. Each will is positioned to slide into one of two grooved rails that are installed flash with the floor.

Single-bolster operation. A rigid chain system pushes the bolster sideways in a straight line out of the press during changeover. This may be done at the front, rear, or sides of the press. At this point an operator can unbolt the die and remove it with an overhead crane. The bolster is then ready to receive the new die (usually introduced via the overhead crane), which is manually clamped to the bolster and pulled back into the press. Finally, the bolster itself is clamped into position.

The whole process can be completed in 10 to 15 minutes.

The advantages of this system are that it provides a means to easily install a die at a preset position and ensure both uniform clamping and uniform positioning in the press. The disadvantage of this system is that the press does not function during the entire time that the die exchange is taking place.

Double-bolster operation. To reduce downtime, this system uses two bolsters that can be moved either side-to-side or front-to-back, although side-to-side motion is preferred. The second bolster allows an operator to pre-stage a new die before the changeover. When the rigid chain is activated, the first bolster is pushed out of the press, and the second bolster, carrying the new die, is pulled in. This process allows for a faster die exchange, reducing the time spent to as little as five minutes. The main advantage of this system is that press downtime is reduced and press utilization improved. The disadvantage to this system is that more space is needed around the press in order to accommodate the bolster that is not in operation. This type of quick die change is also greatly dependent on the press design, although modifications to the press are often possible.

Rolling bolster positioning methods. There are two locating methods commonly used with rolling bolsters. The first and probably most common uses special wedges located on the bed of the presses and on the bolsters. As the bolster is brought into position and lowered, the wedges force the bolster into a correct position. The second method uses index pins on the top of the bolster; locators in the die also assure uniformity. Bolster clamping can be accomplished by manual means, a hydraulically operated swing clamp system, a T-slot clamp system, or a ledge clamp system. The type of clamping system used is mostly dependent on the type of press construction. The upper shoes may be clamped either manually, by inserting hydraulic clamps in the tee slots; by swing clamps located in a fixed location; or by traveling clamps that will adjust automatically to the upper shoe size. Lower die-shoe clamping is normally a manual operation performed when the dies are placed onto the bolster. Hydraulic clamps have been used to reduce the number of tools and skilled tradespeople required for die setting.

b) Die Cart Systems

The primary function of these rail-guided carts is to remove dies from press beds and replace them with new dies removed from storage. The carts can handle dies of more than 100 tons and can reduce changeover time to less than five minutes.

There are many versions of die cart systems, but the most popular die cart systems are the following:

- single-stage die cart on rails
- double-stage die cart on rails
- two single-stage die carts on rails

Single-stage die cart on rails. This system incorporates the following elements: one single-stage die cart on rails, one push-pull system (PPS), and two staging tables. The system is dedicated to one press. Because the cart is combined with the die-staging tables, exchange of a die may commence while the press is still running. Figure 14.9 shows a schematic process drawing of a quick die change using this method:

1. The old die is pushed out of the press and onto the die cart. The cart brings it to the table on the right.
2. The cart now moves to the table on the left, picks up the new die, and brings it to the press, where the die loader pulls it into its position on the press.

This method is typically less expensive than the rolling bolster system and provides the same benefits as the single rolling bolster.

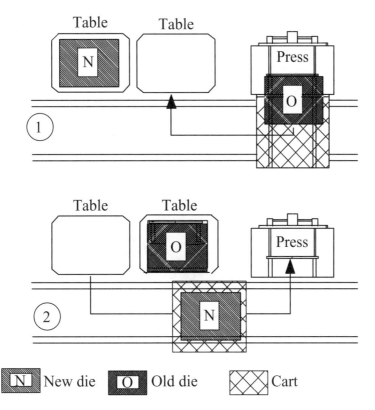

Fig. 14.9 Single-stage die cart on rails.

Dual-stage die cart on rails. A dual-stage die cart on rails incorporates the following elements: one double-stage die cart on rails and one PPS. This system is dedicated to one press, and allows a die to be pre-staged onto one side of the cart in preparation of an upcoming changeover. When the change becomes necessary, a single operator moves the empty side of the cart into position in front of the press. The next steps of the changing process are as shown in Fig. 14.10.

1. An automated push-pull rigid chain system is then activated to pull the old die or sometimes the entry bolster onto the empty cart.
2. The cart is then repositioned on the right to align the new die in front of the press, and the push-pull rigid chain is activated once again to position the die within the press bed. The cart may now take the old die to storage.

A double-stage die cart system provides a very fast die-exchange time of at most three minutes. The main advantage of this system is that downtime of the press is decreased and press utilization factor increased thanks to the very short exchange times, providing an early return on investment.

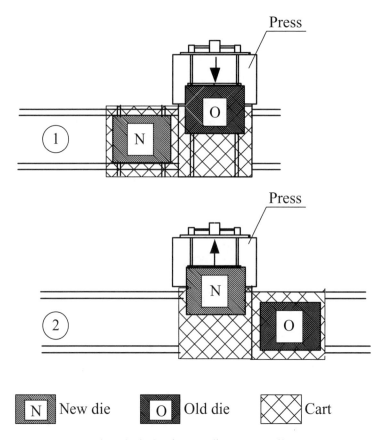

Fig. 14.10 Dual-stage die cart on rails.

Two single-stage die carts on rails. This die transfer equipment typically is mounted to the floor in the rear of the press. These tables resemble the capital letter T or L. Both of these configurations allow two dies to be used simultaneously. A new die can be prepared and positioned to the left or right until the moment for the change-over. Figure 14.11 shows a schematic illustration of this method of quick die change:

1. When the press stops, an automated push-pull rigid chain system is activated to pull the old die out of the press. It then moves backwards on its rails, which run straight to the press.

2. The cart with the new die now moves in front of the press into center position, so that it stands exactly between the press and the cart holding the old die.

3. The push-pull unit on the second cart pushes the new die into the appropriate position on the press bed.

4. The cart that brought the next die now takes the old die back to storage.

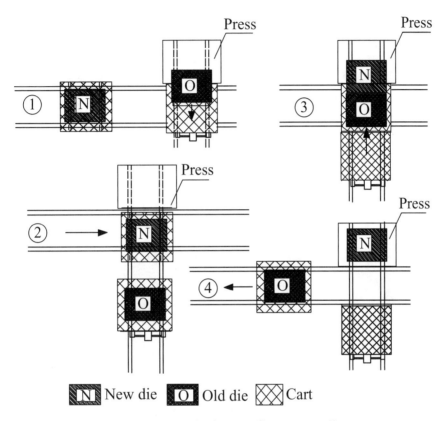

Fig. 14.11 Two single-stage die carts on rails.

This system offers high speed die change and high flexibility. Bolster heights are allowed to vary and the dies may be alternated. Cost and floor space requirements are major disadvantages of the system.

Die cart positioning method. The positioning method generally used for die cart transfer systems consists of index pins mounted in the carts extending into mating sockets in the floor.

14.2.3 Other Components of Quick Die-Change Systems

a) Die Arms

Die arms, also known as bolster extensions, facilitate the motion of dies or subplates to an accessible area of the press for convenient die changing. These units are structural frames with rollers on top that support both the die and the subplate outside of the press, are typically attached to the bolster, and are an integral part of the change-over operation. Bolster extensions can be fixed, temporarily held in place by quick-release details, mounted on a pivot to swing out when not in use, or mobile (mounted on casters so they can be rolled in and out of position).

During changeovers, the old die is manually pulled onto bolster extensions. It is removed via forklift or crane, and the new die is placed on the extensions. The new die is then manually pushed into the press, clamped in place, and production restarts. This system is typically used with dies weighing less than 2500 kg. Bolster extensions are easy to install on the front and rear of a press, and a die can be pre-staged such that it does not interfere with the die being removed. Figure 14.12 shows three types of die arms.

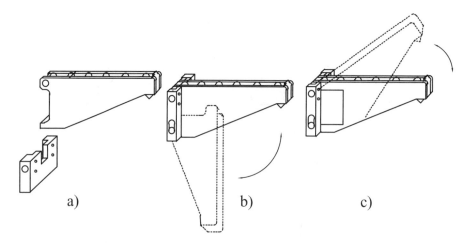

Fig. 14.12 Die arm types: a) detachable type; b) drop-down type; c) folding type.

b) Die Lifter Components

Die lifters are used for simple and quick movement of the dies on the press bed. When dies weigh more than 500 kg, conventional methods of changing dies become heavy, protracted, and often dangerous. Long downtimes result in unproductive manufacturing. By increasing the ease and speed with which die changes can be performed, die lifters increase productivity. These die lifters are spring, hydraulic, or air operated lifting bars. On these bars, either with balls or rollers, the die or subplate can be simply moved into its position on press or outside the press. These die lifters fit the standard T slot dimensions as well as rectangular U shaped slots.

Normally, the die lifters are just below the press bed surface and the die cannot be moved, but after being activated by a spring, by hydraulic pressure, or by air pressure, the die lifter is raised a few mm above the press bed surface, clearing the die from the bed and allowing it to be moved simply and easily.

Spring loaded die lifter. Spring-loaded die lifers are very suitable for small- and mid-size dies. They do not require any power source to be actuated because they are actuated by the spring package underneath the ball. Balls allow more movement of the die than rollers. Because balls have a point connection with the die, the load that they are capable of bearing is less than what comparable rollers with a line connection would be able to support. Load distribution over several balls permits smooth movement and it makes it possible to move the die with little effort in any direction.

By closing and clamping the die, the clamping elements overcome the spring force and clamp the die to the press table. When the die clamps are released, the spring-loaded balls lift the die back into its original position where it can be moved. Figure 14.13 shows ball type spring-loaded die lifters.

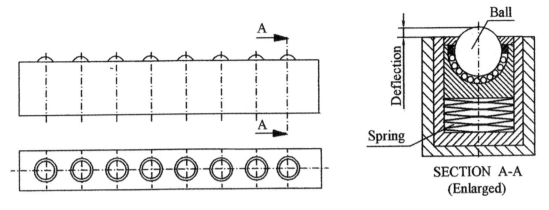

Fig. 14.13 Spring-loaded die lifter.

Hydraulic operated roller die lifters. Hydraulic die lifters are used for heavy dies. The balls and rollers are fitting the bar with each one acting as a hydraulic piston. A pin through the bar prevents them from rotating. When these inserts are pressurized, the balls or rollers make contact with the die and lift the die above the table. When the die is in position, the inserts are depressurized to allow the lowering of the die and then clear the dies bottom surface. The die lifter consists of bars with 3, 4, or 5 inserts. These bars can be connected to each other by means of a connector plug, with a hose connector part added at one end of the linked bars and a plug at the other to prevent hydraulic oil from escaping. This plug is also used to remove air from the system.

14.3 IMPLEMENTATION OF QUICK DIE-CHANGE SYSTEMS

When setting up a quick die-change system, a variety of equipment and process choices, best made by the quick die change team, must be made in order for the system to be effective. The team should include management, manufacturing and tooling engineers, production and maintenance supervisors, setup people, and press operators.

The following factors must be considered when choosing a die changing system:

- What is the design of the dies?
- How large is the available budget?
- What are the present and long-range production requirements?
- What is the goal for the die change time?
- What is the frequency of die changes?
- How many dies are used in each press?
- What are the sizes and weights of the dies involved?
- How many presses are there, and what type of press are they?
- What is the degree of automation?
- How large is the available floor space, and what are the floor conditions?
- What are the minimum and maximum possible press shut heights?
- How far away is the die storage location?
- Is a crane or forklift available?
- What safety hazards must be considered?

The most appropriate method to be used for a given press may be determined by carefully examining all production requirements and related data. Look closely at the press room layout and the presses involved. How many dies are used in each press? What are their size ranges and weights? Examine the present clamping method and clamping points. Review your present method of changeover. Analyze every step and the sequence required to make a die change. How much time is required for each step? Who is involved in the die changing process? What tools and materials are required? The objective is to minimize the steps required and not to duplicate work.

14.3.1 Die Design for Quick Die-Change Systems

The design of the stamping dies is seldom thought of as a means of achieving quick change, and quick change is rarely taken into consideration during die design. Most dies are simply designed according to traditional methods and using conventional wisdom.

The engineering function is most often thought of as the design of the shape and tolerance for proper workpiece and greatest productivity. Designers must bear in mind, however, that their job is to design a die which can produce acceptable parts the first time; that is, without reworking after the dies are in the press. It should be obvious that even the quickest die change will be useless if the dies cannot produce an acceptable part just as quickly. Each interruption to rework some detail or to remove the die for complete rework adds to the change time and wastes production time.

a) New Die Design

When tool and die engineering incorporates die design for QDC systems, the die designers' responsibility is greater than it had been previously, for they are burdened with doing the job right the first time. The discipline of careful record keeping of job successes and failures, thoughtful consideration of reasons for failure and success, diligent documentation of each job, and seeking to continuously improve are all needed to achieve the goal of QDC.

The designers' job is not complete with the design of a functional die. They must also estimate the stock size. If the stock size is too small, the result will be either scrap parts or time lost in waiting for the proper material to arrive. If the stock is too large, die wear and scrap costs will increase due to excess material being discarded in the form of flash. Costs will increase and the advantages of quick die change may be lost through inefficiency. The designer must strive to have all dies designed with the same clamping height. Standardization allows use of the same size of clamping fixture for all dies. Furthermore, if die size is standardized, then fixed die-locating pins can be used in the press bed for quick, accurate die locating. Clamp placement can be easier, or possibly even automatic. If it is not possible to design all dies with same standard dimensions, a few standard sizes covering the complete range of dies may be developed.

b) Accommodating Older Dies in a QDC System

When the time comes to consider equipment for a quick die-change system, it soon becomes clear that most older dies were not designed for standardization. A common problem associated with retrofitting die change equipment is interfacing existing dies with the new system. These existing dies can be all sizes, mounted with or without parallels, and require different clamping methods. A painless way to overcome these varying parameters is to add a subplate of standard thickness and standard dimensions.

The subplate acts as a common dieholder to maintain system uniformity. Subplates are normally the same size as the bolster, so die size variation is not important as long as it does not exceed the bolster's size. The locating subplate in the press can be simple, using locating pins or die guides on the press bolster to allow consistently quick and accurate die setting. Furthermore, clamping it to the bolster is consistent and fast because the subplate need not be changed. When the subplate is located accurately in the press and the die is located accurately on the subplate, repeatable die location is ensured every time. This saves startup time with any associated automation such as coil feeders, pick-and-place units, and transfers. The subplate provides a flat surface that contacts rollers in the press and other equipment that handles the die during the changeover process. This minimizes or eliminates the cost of having to alter dies so that they interface with the die-change system. An added bonus is that the subplate covers the bolster, keeping it free of slugs and debris.

14.3.2 Selection of Die Transfer Methods

Choosing die transfer equipment is based on space needs, the desired die change time, the available budget, the frequency of die changes, and the desired degree of automation. There are three basic die options for all die sizes:

- rolling bolster systems
- die cart systems
- die arms bolster extensions

Rolling bolster systems and die cart systems are more expensive, but when these types of equipment are used as a part of QDC systems, changing time may be reduced to between 3 and 15 minutes. Furthermore, these systems allow for a high degree of automation, offer high flexibility, and yield a high safety environment. A major disadvantage, however, is that cost and space requirements for both systems are very high, particularly for die cart systems.

The die arms (bolster extensions) transform system is usually employed for smaller die sizes. This system is typically used for dies weighing less than 2500 kg. Major advantages of bolster extension are that equipment is easy to install on the front and rear of a press, a die can be pre-staged without interfering with the die being removed, and the cost of the system is lower than for die cart systems and die arms bolster extensions. Disadvantages of this system are a die changing time of 15 to 20 minutes and reduced safety.

14.3.3 Selection of Die Clamping Systems

To facilitate choosing quick clamping and unclamping systems, standardize the clamping method and clamping height for a particular press or group of presses. Grouping presses by size and die type is the best way to break down clamping equipment into common components. With standardized clamping heights, shut heights, and location of clamps, clamping procedures may be rendered simple and quick, regardless of the chosen clamping system.

When a clamping system is implemented on the press, it should be ensured that the clamps will not interfere with die operation and can be moved for die removal. Generally, clamps are removed from the slots before a die is inserted or removed. However, if the die size or press arrangement prevents removal, it may be desirable to slide the dies in under the clamps. In any case, the ram and bolster plates need to have the necessary number of slots and locations for the clamps to reach all the dies in the press, as well as to clear the dies during installation and removal.

a) Calculate Clamping Force

Choose the clamp force rating needed to exceed the clamping requirements of the most demanding die. Clamp force must be calculated based on the maximum die weight that will be run in the press, using the clamp force rating needed to exceed the clamping requirements of the most demanding die. Clamp force is calculated for the upper clamps as follows:

$$F_c = \frac{W_{d(max)} \times 4}{N_c}$$
(14.1)

where

F_c = clamp force per clamp

$W_{d(max)}$ = maximum die weight

N_c = number of clamps for upper clamping

The press should be balanced using an equal number of bottom and upper clamps. Large clamp forces are needed because the press movement causes dynamic loading. Note that these calculations are general recommendations for moderate press speeds and should not be used to determine a particular press's requirements without consulting manufacturer recommendations and internal company policy. To develop a plant standard, consider all the data from several presses when possible, rather than from a single press.

14.3.4 Selection of Die Lifter System

The concentration of die mass per length unit of the die lifter (also known as linear die density) determines the selection of lifting device. Low-density dies generally can be lifted with ball roller rails, spring-loaded ball rails, or ball cartridges, which allow the die to be moved in any direction. High-density dies usually require several ball-style rails or a pair of hydraulic operated roller die lifters, which move in only one direction. If a die needs to be moved in both the inline and transverse direction, a pair of insertion rails and a pair of transverse rails is required. Generally, it is recommended to plan for both directions of movement to take full advantage of the QDC equipment's ability to index the die's exact location and alignment prior to clamping.

a) Calculation of Lifter Capacity

In order to calculate lifter capacity, it is necessary to know the die length in the direction of the slots (assuming that the slots are in the same direction as die insertion), and the weight of each of the dies. Calculate die density as follows:

$$D_d = \frac{W_d}{D_L}$$
(14.2)

where

D_d = die density, kg/m (lb/ft)

W_d = die weight, kg (lb)

D_L = die length in the direction of the slots, m (ft)

Select a die lifter with enough capacity to lift the die weight and move the die in the desired direction. It is very important that the die lifter is not chosen based on overall length, as only the length of the die lifter that is under the die at any given time is actually lifting the die.

Once the capacity of the die lifter needed for the densest die has been calculated, confirm that the lifters chosen have the capacity to lift all other dies. Typically, choosing the densest die will ensure the correct capacity. However, slot location can be a problem if both large and small dies are run in the same press, because some of the dies might be too small to have all the rails under them.

Perform lifter calculations for all dies to verify that each die can be lifted, based on the requirement that they all satisfy the following relationship:

$$W_d < (D_L \times L_d \times N_L) \tag{14.3}$$

where

W_d = die weight, kg (lb)

D_L = die length in the direction of the slots, m (ft)

L_d = lifter density, kg/m (lb/ft)

N_L = number of lifters under the specific die

14.3.5 Selection of Support Equipment

Consider the activating and controlling equipment (pumps, hoses, manifolds, mounting devices, for the clamps, lifters, transferring systems, and the control device) as well as the space required to mount them. Generally, the most attention must be paid to the selection of permanently mounted pumps and the manual valves to operate the clamps. Fixed hydraulic equipment is typically attached to presses to facilitate QDC installations. More recently, electronic valve controllers tied into the press control or at the press control via keyed switches are being used instead.

14.3.6 Safety Circuits and Evaluation

After the QDC system parameters have been determined, safety circuits and integration into press controls must also be reviewed. Separate pilot-operated check valves for each clamp provide a very high safety level. Check valves providing a dual diagonal hydraulic safety circuit also ensure safe clamping for either the press bed or the slide plates.

Depending on the level of automation desired, different electrical controls to press and clamp controls may be required. The type of signal is determined by the pressure switch, clamp position, die position, slide plate position, continuous/run setup, and the enabling of the press to operate.

a) Evaluation

Once a suitable die-change system has been installed, it must be evaluated by the quick die change team. Are the new principles for die change and movement providing the desired results? Is the clamp system safe, effective, and reliable? Can the new die change time be further improved?

Based on labor savings, increased press utilization, lowered inventories, reduced scrap, and increased safety, project the cost savings for a one-year period. If *payback* is as expected, it is time to repeat the process and move on to the next press or presses.

$$t = \frac{(CI \times 14.4)}{DC \times M \times t_s} \tag{14.4}$$

where

t = payback time in months

CI = capital investment

DC = die change per week

M = hourly machine cost

t_s = time saved in minutes by the QDC system

Example:

Calculate the payback time in months if the total capital investment for quick die change equipment is $CI = \$\,25,000$; die change per week $DC = 20$; hourly machine cost $M = \$\,75$; and time saving per die change by QDC system $t_s = 20$ minutes.

Solution:

$$t = \frac{(CI \times 14.4)}{DC \times M \times t_s} = \frac{25,000 \times 14.4}{20 \times 75 \times 20} = 12 \text{ months}$$

Note: This calculation does not include labor-saving measures, reduction of inventory, increased safety, or improved working conditions.

Appendix
One

Blank Diameter of Drawn Shells

In the accompanying diagrams, it should be noted that the solid line is the mean line of the drawn part. This line and its associated dimension should be used when calculating the diameter for drawn shells. Table A1.1 provides sample equations for blank-diameter calculation of symmetrical shells.

Table A1.1 Equations for calculating blank-diameters of drawn shells

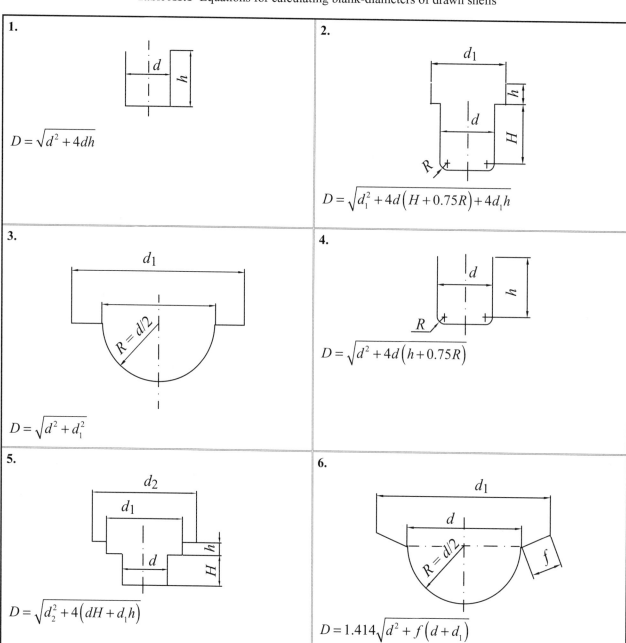

1.

$$D = \sqrt{d^2 + 4dh}$$

2.

$$D = \sqrt{d_1^2 + 4d(H + 0.75R) + 4d_1 h}$$

3.

$$D = \sqrt{d^2 + d_1^2}$$

4.

$$D = \sqrt{d^2 + 4d(h + 0.75R)}$$

5.

$$D = \sqrt{d_2^2 + 4(dH + d_1 h)}$$

6.

$$D = 1.414\sqrt{d^2 + f(d + d_1)}$$

7.

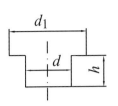

$$D = \sqrt{d_1^2 - 4dh}$$

8.

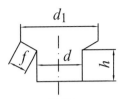

$$D = \sqrt{d^2 + 4dh + 2f\left(d + d_1\right)}$$

9.

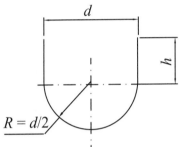

$$R = d/2$$

$$D = \sqrt{8R^2 + 4dh}$$

10.

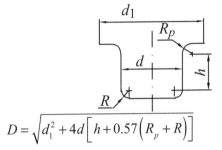

$$D = \sqrt{d_1^2 + 4d\left[h + 0.57\left(R_p + R\right)\right]}$$

11.

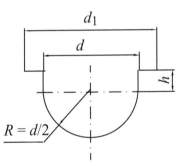

$$R = d/2$$

$$D = \sqrt{d^2 + d_1^2 + 4dh}$$

12.

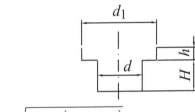

$$D = \sqrt{d_1^2 + 4\left(dH + d_1h\right)}$$

13.

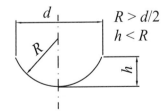

$R > d/2$
$h < R$

$$D = \sqrt{d^2 + 4h^2}$$

14.

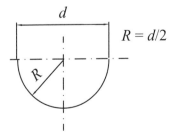

$R = d/2$

$$D = \sqrt{8R^2}$$

15.

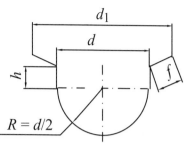

$$D = 1.41\sqrt{d^2 + 2dh + f(d + d_1)}$$

16.

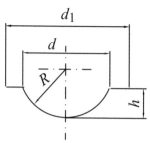

$$D = \sqrt{d_1^2 + 4h^2}$$

17.

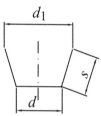

$$D = \sqrt{d + 2s(d + d_1)}$$

18.

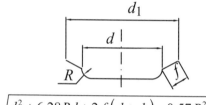

$$D = \sqrt{d^2 + 6.28Rd + 2f(d + d_1) - 0.57R^2}$$

19.

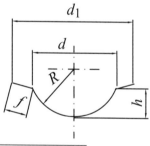

$$D = \sqrt{d^2 + 4h^2 + 2f(d + d_1)}$$

20.

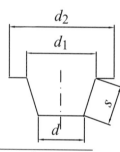

$$D = \sqrt{d^2 + 2s(d + d_1) + d_2^2 - d_1^2}$$

21.

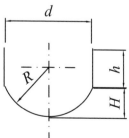

$$D = \sqrt{d^2 + 4(H^2 + dh)}$$

22.

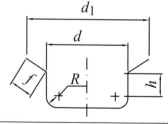

$$D = \sqrt{d^2 + 4d\left(0.57R + h + \frac{1}{2}\right) + 2d_1 f - 0.57R^2}$$

23.

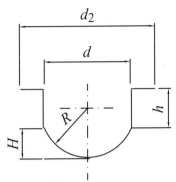

$$D = \sqrt{d_1^2 + 4\left(H^2 + dh\right)}$$

24.

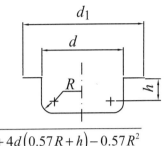

$$D = \sqrt{d_1^2 + 4d\left(0.57R + h\right) - 0.57R^2}$$

25.

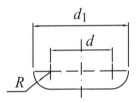

$$D = \sqrt{d^2 + 6.28Rd + 8R^2}$$

26.

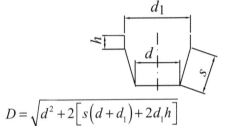

$$D = \sqrt{d^2 + 2\left[s\left(d + d_1\right) + 2d_1h\right]}$$

27.

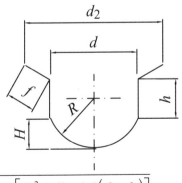

$$D = \sqrt{d^2 + 4\left[H^2 + dh + 0.5\left(d + d_2\right)\right]}$$

28.

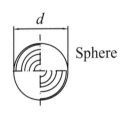

Sphere

$$D = 1.128\sqrt{\pi d^2} = \sqrt{4d^2} = 2d$$

Appendix
Two

Metric System Tolerances On Linear Dimensions

A2.1 DEFINITIONS

In the metric system, several terms are used to describe features of dimensional relationships between mating parts:

Dimensional tolerance. Dimensional tolerance is defined as the permissible or acceptable variation in the dimensions of a part.

Nominal size. The dimension that is used for the purpose of general identification. It is written into drawings and other technical documentation.

True size. The dimension that is measured on a finished part. Like all other sizes, this size includes inaccuracy of measuring.

Limit sizes. Limit sizes are two given sizes, between which must be the true size of a correct piece.

Upper limit size. Upper limit size is the maximum allowance for the dimension of a correctly made piece.

Lower limit size. Lower limit size is the minimum allowance dimension of a correctly made piece.

Maximum material condition. A piece whose dimensions are at the upper limit size.

Minimum material condition. A piece whose dimensions are at the lower limit size.

Deviation. Algebraic difference between some certain size and nominal size. The value is positive if this size is larger than nominal size and negative if it is less than nominal size.

Zero line. A straight line that, in a graphical interpretation of tolerance, corresponds with nominal size, so it is the beginning line for the calculation of deviations.

Upper deviation. Algebraic difference between upper limit size and nominal size.

Lower deviation. Algebraic difference between lower limit size and nominal size.

True deviation. Algebraic difference between true size and nominal size.

Tolerance. Tolerance is the algebraic difference between upper limit size and lower limit size.

Fit. The relationship between mating parts of the same nominal size, which stem from the differences in the true sizes before assembling.

Clearance. The relationship between mating parts of the same nominal size that stems from the difference in their true sizes if the true size of the hole is bigger than the true size of the shaft before assembling. Accordingly, the clearance is always positive.

Clearance fit. Fit that allows for rotation or sliding between mating parts.

Interference. The relationship between mating parts of the same nominal size that stems from differences in their true sizes: if the true size of the hole is smaller than the true size of the shaft, before assembling, the clearance is always negative.

Interference fit. A fit that everywhere provides interference between the hole and the shaft when assembled, i.e., the maximum size of the hole is smaller than the minimum size of the shaft.

Transition fit. A fit that may provide either a clearance or interference between the hole and the shaft when assembled, depending on whether the hole and shaft overlap completely or in part.

Hole-basis system. Tolerances are based on a zero line on the hole; also called the basic hole system. The tolerance position with respect to the holes is always represented by the capital letter *H* (lower limit size equals zero). The desired fit is achieved by selecting the correct tolerance position of the mating shaft (indicated by a lower-case letter) of the same nominal size.

Shaft-basis system. Tolerances are based on a zero line on the shaft; also called the basic shaft system. The tolerance position is always represented by the lower-case letter *h* (upper limit size equals zero). The desired fit is achieved by selecting the correct tolerance position of the mating hole (indicated by a capital letter) of the same nominal size.

Figure A2.1 shows a graphical interpretation of nominal size, limit size, deviation, and tolerance on shaft and hole. Figure A2.2 shows tolerance positions with regard to the zero line.

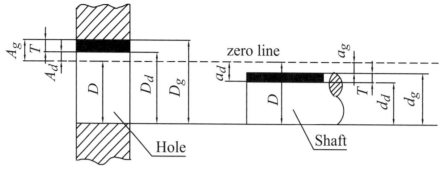

LEGEND:
D- nominal size
d_d- lower limitof shaft size
d_g- upper limit of shaft size
a_d- lower deviation of shaft
a_g- upper deviation of shaft

D_d- lower limit of hole size
D_g- upper limit of hole size
A_d- lower deviation of hole
A_g- upper deviation of hole
T- tolerance

Fig. A2.1 Limit sizes, deviations, and tolerance.

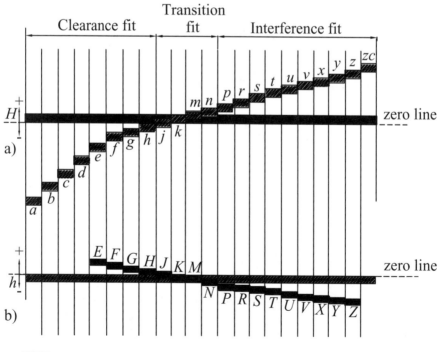

■ Tolerance position of the hole-basis system
▨ Tolerance position of the shaft-basis system

Fig. A2.2 Schematic illustration of tolerance positions: a) hole-basis system; b) shaft-basis system.

A2.2 SYSTEM OF LIMITS AND FITS

Tolerances in the metric system are designated by the nominal size in millimeters, the letter designation of the tolerance position with regard to the zero line, and the grade of the tolerance position.

Example:

$$\phi \ 50 \ a11, \ \ \phi \ 100 \ H8.$$

Designation of fits in the metric system of tolerances is expressed as the nominal size (mm), followed by the tolerance position for the hole over the tolerance position for the shaft.

Example:

$$\phi \ 50 \ H6/s5 \ \ or \ \ \phi \ 30 \ \frac{P6}{h7}$$

Table A2.1 gives an example of the metric system fits classified by the International Standards Organization (ISO) using the hole-basis system. Table A2.2 gives the most-used metric tolerances for holes. Table A2.3 gives the most-used tolerances for shafts.

Table A2.1 Preferred fits using the hole-basis system

Fit classifications	Description	Hole-basis	Directive for using
Clearance fits	Loose running	H11/c11; C11/h11	Fit for commercial tolerance or allowances on external members.
	Free running	H9/d9; D9/h9	This fit is not for use where accuracy is essential, but is good for large temperature variations, high running speeds, or heavy journal pressures.
	Close running	H8/f7; F8/h7	Fit for running on accurate machines and for accurate locations at moderate speeds and journal pressures.
	Sliding	H7/g6; G7/h6	Fit not intended to run freely, but to move and turn freely and locate accurately.
	Location clearance	H7/h6; H6/g5	Fit provides snug seat for locating stationary parts, but can be freely assembled and disassembled.
Transition fits	Location-slight interference	H7/k6; K7/h6	Fit for location, a compromise between clearance and interference.
	Location/ Transition	H7/n6; N7/h6	Fit for more accurate location where greater interference is permissible.
Interference fits	Location interference	H7/p6; P7/h6	Fit for parts requiring rigidity and alignment with prime accuracy of location but without special bore pressure requirements.
	Medium drive	H7/s6; S7/h6	Fit for ordinary steel parts or shrink fits on light sections, the tightest fit usable with cast iron.
	Force	H7/u6; U7/h6	Fit suitable for parts, which can be highly stressed or for shrink fits where the heavy pressing forces required are impractical.

Table A2.2 Metric system tolerance for holes

						Nominal hole size (mm)									
over	3	6	10	18	30	40	50	65	80	100	120	140	160	180	200
inc.	6	10	18	30	40	50	65	80	100	120	140	160	180	200	225
Micrometers															
E6	+28 +20	+34 +25	+43 +32	+53 +40	+66 +50	+79 +60			+94 +72		+110 +85			+129 +100	
E7	+32 +20	+40 +25	+50 +32	+61 +40	+75 +50	+90 +60			+107 +72		+125 +55			+146 +100	
E11	+95 +20	+115 +25	+142 +32	+170 +40	+210 +50	+250 +60			+292 +72		+335 +85			+390 +100	
E12	+14 0 +20	+175 +25	+212 +32	+250 +40	+300 +50	+360 +60			+422 +72		+485 +85			+560 +100	
E13	+20 0 +20	+245 +25	+302 +32	+370 +40	+440 +50	+520 +60			+612 +72		+715 +85			+820 +100	
F6	+18 +10	+22 +13	+27 +16	+33 +20	+41 +2	+49 +30			+58 +36		+68 +43			+79 +50	
F7	+22 +10	+28 +13	+34 +16	+41 +20	+50 +25	+60 +30			+71 +36		+68 +43			+96 +50	
F8	+28 +10	+35 +13	+43 +16	+53 +20	+64 +25	+76 +30			+90 +36		+106 +43			+122 +50	
G6	+12 +4	+14 +5	+17 +6	+20 +7	+25 +9	+29 +10			+34 +12		+39 +14			+44 +15	
G7	+16 +4	+20 +5	+24 +6	+28 +7	+34 +9	+40 +10			+47 +12		+54 +14			+61 +15	
G8	+22 +4	+27 +5	+33 +6	+40 +7	+48 +9	+56 +10			+66 +12		+77 +14			+87 +15	
H6	+8 0	+9 0	+11 0	+13 0	+16 0	+19 0			+22 0		+25 0			+29 0	
H7	+12 0	+15 0	+18 0	+21 0	+25 0	+30 0			+35 0		+40 0			+46 0	
H8	+18 0	+22 0	+27 0	+33 0	+39 0	+46 0			+54 0		+63 0			+72 0	
H9	+30 0	+36 0	+43 0	+52 0	+62 0	+74 0			+97 0		+100 0			+115 0	
H10	+48 0	+58 0	+70 0	+84 0	+100 0	120 0			+140 0		+160 0			+185 0	

	+75 / 0	+90 / 0	+110 / 0	+130 / 0	+160 / 0	+190 / 0			+220 / 0		+250 / 0			+290 / 0
H11	+75 / 0	+90 / 0	+110 / 0	+130 / 0	+160 / 0	+190 / 0			+220 / 0		+250 / 0			+290 / 0
J6	+5 / −3	+5 / −4	+6 / −5	+8 / −5	+10 / −6	+13 / −6			+16 / −6		+18 / −7			+22 / −7
J7	+6 / −6	+8 / −7	+10 / −8	+12 / −9	+14 / −11	+18 / −12			+22 / −13		+26 / −14			+30 / −16
J8	+10 / −8	+12 / −10	15 / −12	+20 / −13	+24 / −15	+28 / −18			+34 / −20		+41 / −22			+47 / −25
K6	+2 / −6	+2 / −7	+2 / −9	+2 / −11	+3 / −13	+4 / −15			+4 / −18		+4 / −21			+5 / −24
K7	+3 / −9	+5 / −10	+6 / −12	+6 / −15	+7 / −18	+9 / −21			+10 / −25		+12 / −28			+13 / −33
K8	+5 / −13	+6 / −16	+8 / −19	+10 / −23	+12 / −27	+14 / −32			+16 / −38		+20−43			+22 / −50
M6	−1 / −9	−3 / −12	−4 / −15	−4 / −17	−4 / −20	−5 / −24			−6 / −28		−8 / −33			−8 / −37
M7	0 / −12	0 / −15	0 / −18	0 / −21	0 / −25	0 / −30			0 / −35		0 / −40			0 / −46
M8	+2 / −16	+1 / −21	+2 / −25	+4 / −29	+5 / −34	+5 / −41			+6 / −48		+8 / −55			+9 / −63
N6	−5 / −13	−7 / −16	−9 / −20	−11 / −24	−12 / −28	−14 / −33			−16 / −38		−20−45			−22 / −51
N7	−4 / −16	−4 / −19	−5 / −23	−7 / −28	−8 / −33	−9 / −39			−1- / −45		−12 / −52			−14 / −60
N8	−2 / −20	−3 / −25	−3 / −30	−3 / −36	−3 / −42	−4 / −50			−4 / −58		−4 / −67			−5 / −77
P6	−9 / −17	−12 / −21	−15 / −26	−18 / −31	−21 / −37	−26 / −45			−30 / −52		−36 / −61			−41 / −70
P7	−8 / −20	−9 / −24	−11 / −29	−14 / −35	−17 / −42	21 / −51			−24 / −59		−28 / −68			−33 / −79
P8	−12 / −30	−15 / −37	−18 / −45	−22 / −55	−26 / −65	−32 / −78			37 / −91		−43 / −106			−50 / −122
R6	−12 / −20	−16 / −25	−20 / −31	−24 / 37	29 / 45	−35 / −54	−37 / −56	−44 / −66	−47 / −69	−56 / −81	−58 / −83	−61 / −86	−68 / −97	−71 / −100
R7	−11 / −23	−13 / −28	−16 / −34	−20 / −41	−25 / −50	−30 / −60	−32 / −62	−38 / −73	−41 / −76	48 / −88	−50 / −90	−53 / −93	−60 / −106	−63 / −109

Table A2.3 Metric system tolerance for shafts

Nominal shaft size (mm)															
over	3	6	10	18	30	40	50	65	80	100	120	140	160	180	200
inc.	6	10	18	30	40	50	65	80	100	120	140	160	180	200	225
micrometers															
a12	−270 −390	−280 −430	−290 −470	−340 −640	−340 −640	−340 −640	−340 −640	−360 −660	−520 −920	−520 −920	−520 −920	−520 −920	−580 −980	−660 −1120	−740 − 120 0
c11	−70 −145	−80 −170	−95 −205	−110 −240	−120 −280	−130 −290	−140 −330	−150 −340	−170 −390	−180 −400	−20 −450	−210 −460	−230 −480	240 −530	−260 −550
d6	−30 −38	−25 −34	−32 −43	−40 −53	−80 −96		−100 −119		−170 −199		−170 −199			−170 −199	
e6	−20 −28	−25 −34	−32 −43	−40 −53	−50 −66		−60 −79		−72 −94		−85 −110			−100 −129	
e13	−20 − 200	−25 − 245	−32 − 302	−40 − 370	−50 −440		−60 −520		−72 −612		−85 −715			−11 −820	
f5	−10 −15	−13 −19	−16 −24	−20 −29	−25 −36		−30 −43		−36 −51		−43 −61			5− −70	
f6	−10 −18	−13 −22	−16 −27	−20 −33	−25 −41		−30 −49		−36 −58		−43 −61			50 −70	
f7	−10 −22	−13 −28	−16 −34	−20 −41	−25 −50		−30 −60		−36 −71		−43 −83			−50 −96	
g5	−4 −9	−5 −11	−6 −14	−7 −−16	−9 −20		10 −23		12 −23		12 −27			−14 −32	
g6	−4 −12	−5 −14	−6 −17	−7 −20	−9 −25		−10 −29		−12 −34		−14 −39			−15 −44	
g7	−4 −16	−5 −20	−6 −24	−7 −28	− −34		−10 −40		−12 −47		−14 −54			−15 −61	
h5	0 −5	0 −6	0 −8	0 −9	0 −11		0 −13		0 −15		0 −18			0 −20	
h6	0 −8	0 −9	0 −11	0 −13	0 −16		0 −19		0 −22		0 −25			0 −29	
h7	0 −12	0 −15	0 −18	0 −21	0 −25		0 −30		0 −35		0 −40			0 −46	

h8	0 / −18	0 / −22	0 / −27	0 / −33	0 / −39	0 / −46		0 / −54			0 / −63		0 / −72
h10	0 / −48	0 / −58	0 / −70	0 / −84	0 / −100	0 / −120		0 / −140			0 / −160		0 / −185
h11	0 / −75	0 / −90	0 / −110	0 / −130	0 / −160	0 / −190		0 / −220			0 / −250		0 / −290
j5	+3 / −2	+4 / −2	+5 / −3	+5 / −4	+6 / −5	+6 / −7		+6 / −9			+7 / −11		+7 / −13
j6	+6 / −2	+7 / −2	+8 / −3	+9 / −4	+11 / −5	+12 / −7		13 / −9			14 / −11		+16 / −13
j7	+8 / −4	+10 / −5	+12 / −6	+13 / −8	+15 / −10	+18 / −12		+20 / −15			+22 / −18		+25 / −21
k5	+6 / +1	+7 / +1	+9 / +1	+11 / +2	+13 / +2	+15 / +2		+18 / +3			+21 / +3		+24 / +4
k6	+9 / +1	+10 / +1	+12 / +1	+15 / +2	+18 / +2	+21 / +2		+25 / +3			+28 / +3		33 / +4
k7	+13 / +1	+16 / +1	+19 / +1	+23 / +2	+27 / +2	+32 / +2		+38 / +3			+43 / +3		+50 / +4
m5	+9 / +4	+12 / +6	+15 / +7	+17 / +8	+20 / +9	+24 / +11		+28 / +13			+33 / +15		+37 / +17
m6	+12 / +4	+15 / +6	+18 / +7	+21 / +8	+25 / +9	+30 / +11		+35 / +13			+40 / +15		+46 / +17
m7	+16 / +4	+21 / +6	25 / +7	+29 / +8	+34 / +9	+41 / +11		48 / +13			+55 / +15		+63 / +17
n5	+13 / +8	+16 / +10	+20 / +12	+24 / +15	+28 / +17	33 / +20		38 / +23			45 / +27		+51 / +31
n6	+16 / +8	+19 / +10	23 / +12	+28 / +15	+33 / +17	+39 / +20		+45 / +23			+45 / +27		+51 / +31
n7	+20 / +8	+25 / +10	+30 / +12	+36 / +15	+42 / +17	+50 / +20		+58 / +23			+67 / +27		+77 / +31
p5	+17 / +12	+21 / +15	+26 / +18	+31 / +22	+37 / +26	+45 / +32		+52 / +37			+61 / +43		+70 / +50
p6	+20 / +12	+24 / +15	+29 / +18	+35 / +22	42 / +26	+51 / +32		+59 / +37			+68 / +43		+79 / +50
r6	+23 / +15	+28 / +19	+34 / +23	+41 / +28	+50 / +34	+60 / +41	+62 / +43	+73 / +51	+76 / +54	+88 / +63	+90 / +65	+93 / +68	+105 / +77	109 / +80
r7	+27 / +15	+34 / +19	+41 / +23	+49 / +28	+59 / +34	+71 / +41	+73 / +43	+86 / +51	+89 / +54	103 / +63	105 / +65	108 / +68	+123 / +77	126 / +80

A2.3 SYSTEM OF GEOMETRIC TOLERANCES

In manufacturing, geometric tolerances are used to specify the maximum allowable deviation from the exact size and shape specified by designers.

Geometric tolerances provide a comprehensive method of specifying where the critical surfaces are on a part, how they relate to one another, and how the part should be inspected to determine if the tolerance is acceptable.

Table A2.4 gives metric standard symbols for geometric tolerances.

Table A2.4 ISO standard system for geometric tolerances

Geometric Characteristic	Symbol	Definitions
Straightness	⎯⎯	**Straightness** All points on the indicated surface or axis must lie in a straight line in the direction shown, within the specified tolerance zone.
Flatness	▱	
Circularity	○	**Flatness** All points on the indicated surface must lie in a single plane, within the specified tolerance zone.
Cylindricity	⌭	
Profile of a Line	⌒	**Circularity** If the indicated surface were sliced by any plane perpedicular to its axis, the resulting outline must be a perfect circle, within the specified tolerance zone.
Profile of a Surface	⌓	
Angularity	∠	
Perpendicularity	⊥	**Cylindricity** All points on the indicated surface must lie in a perfect cylinder around a center axis, within the specified tolerance zone.
Parallelism	∥	
Position	⌖	**Linear Profile** All points on any full slice of the indicated surface must lie on its theoretical two-dimensional profile, as defined by basic dimensions, within the specified tolerance zone. The profile may or may not be oriented with respect to datums.
Concentricity/Coaxiality	◎	
Symmetry	═	
Circular Runout	↗	
Total Runout	↗↗	**Surface Profile** All points on the indicated surface must lie on its theoretical three-dimensional profile, as defined by basic dimensions, within the specified tolerance zone. The profile may or may not be oriented with respect to datums.
At Max. Material Condition	Ⓜ	
At Least Material Condition	Ⓛ	
Regardless of Feature Size	Ⓢ	**Datum Feature** A triangle which designates a physical feature of the part to be used as a reference to measure geometric characteristics of other part features.
Projected Tolerance Zone	Ⓟ	
Diameter	⌀	

Basic Dimension	55
Reference Dimension	(40)
Datum Feature	A1
Target Point	X
Dimension Origin	⊕→
Feature Control Frame	⌒ 0.2
Conical Taper	⊳
Slope	◁
Square (Shape)	□
Dimension Not to Scale	20
Number of Times/Places	5x
Arc Length	⌒100
Radius	R
Spherical Radius	SR
Spherical Diameter	Sφ

Maximum Material Condition

A tolerance modifier that applies the stated tight tolerance zone only while the part theoretically contains the maximum amount of material permitted within its dimensional limits allowing more variation under normal conditions.

There are five types of geometric tolerance:

Form tolerance. How far an actual surface or feature is permitted to vary from the desired form specified by the drawing: flatness, straightness, circularity (roundness), and cylindricity.

Profile tolerance. How far an actual surface or feature is permitted to vary from the desired form on the drawing: profile of a line and profile of a surface.

Orientation tolerance: How far a surface or feature is permitted to vary relative to a datum or datums: covering angularity, perpendicularity, and parallelism.

Location tolerance. How far a surface is permitted to vary from the perfect location implied by the drawings as related to a datum, or other features; for position, concentricity, and symmetry.

Runout tolerance. How far a surface or feature is permitted to vary from the desired form implied by the drawing with full 360° rotation of the part on a datum axis: covers circular runout and total runout.

Appendix
Three

Miscellaneous Information

This appendix offers readers many references for subjects related to the sheet metal forming processes, die design, and the die making trade — these references are not easily found in one place.

A3.1 SHEET METAL

Sheet metal is thin, flat metal mass rolled from slab or ingot; it is typically provided in widths from 24 to 80 inches (600 to 2000 mm), and thicknesses between 0.05 inch and 0.250 inches (1.25 to 6.25 mm). Sheet metal can be supplied in cut-to-length pieces or rolled into large, heavy coils.

A3.1.1 ASTM Standards for Steel and Non-Ferrous Sheet and Strip

This group of ASTM standards covers sheets, coils, and strips for steel and non-ferrous metals used in various applications.

Table A3.1 provides a list of ASTM standards specification and a description of each. This table can help engineers, technicians, and other users of this book find the correct numbers.

A3.1.2 Sheet and Strip Thickness

Table A3.2 gives nominal thicknesses of sheet and strip in gauge, decimal inches, and millimeters, for steel and non-ferrous materials. Note that sheet and strip thicknesses for non-ferrous material are not the same as for ferrous materials for the same gauge number.

Table A3.1 Group of ASTM standards for sheet, and strip

ASTM Specifications	DESCRIPTION
	Steel sheet, and strip
A1011/A1011M	Specification for Sheet and Strip, Hot-Rolled, Carbon, Structural, High–Strength Low-Alloy, and High-Strength Low-Alloy, with Improved Formability.
A109/A109M-98a	Specification for Steel, Strip, Carbon (0.25%max.), Cold Rolled.
A263-94a (1999)	Specification for Corrosion-Resisting, Chromium, Steel-Clad Plate, Sheet, and Strip.
A366/A366M-97e1	Specification for Commercial Steel (CS) Sheet, Carbon (015% max) Cold-Rolled.
A414/A414M-00	Specification for Steel, Sheet, Carbon, for Pressure Vessels.
A424-00	Specification for Steel, Sheet, for Porcelain Enameling.
A505-00	Specification for Steel, Sheet and Strip, Alloy, Hot-Rolled and Cold-Rolled.
A506-93 (1998)	Specification for Steel, Sheet and Strip, Alloy, Hot-Rolled and Cold-Rolled, Regular Quality and Structural Quality.
A507-93 (1998)	Specification for Steel, Sheet and Strip, Alloy, Hot-Rolled and Cold-Rolled, Drawing Quality.
A568/A568M-00	Specification for Steel, Sheet, Carbon, and High-Strength, Low-Alloy, Hot-Rolled and Cold-Rolled.
A606-98	Specification for Steel, Sheet, and Strip, High-Strength, Low-Alloy, Hot-Rolled and Cold-Rolled, with Improved Atmospheric Corrosion Resistance.
A607-98	Specification for Steel, Sheet, and Strip, High-Strength, Low-Alloy, Columbium or Vanadium, or Both, Hot-Rolled and Cold-Rolled.
A611-97	Specification for Structural Steel (SS), Sheet, Carbon, Cold-Rolled.
A620/A620M-97	Specification for Drawing Steel (DS), Sheet, Carbon, Cold-Rolled
A622/A622M-97	Specification for Drawing Steel (DS), Sheet and Strip, Carbon, Hot-Rolled
A635/A635M-98	Specification for Steel, Sheet and Strip, Heavy-Thickness Coils, Carbon, Hot-Rolled.
A659/A659M-97	Specification for Commercial Steel (CS), Sheet and Strip, Carbon (0.16 to 0.25 %max), Hot-Rolled.
A682/A682M-98a	Specification for Steel, Strip, High-Carbon, Cold-Rolled.
A684/A684M-86 (1998)	Specification for Steel, Strip, High-Carbon, Cold-Rolled.
	Stainless Steel Sheet and Strip
A167-99	Specification for Stainless and Hot-Resisting Chromium-Nickel Steel Plate, Sheet, and Strip.
A176-99	Specification for Stainless and Hot-Resisting Chromium Steel Plate, Sheet, and Strip.

A240/A240M-00	Specification for Stainless and Hot-Resisting Chromium-Nickel Steel Plate, Sheet, and Strip for Pressure Vessels.
A264-94a (1999)	Specification for Stainless Chromium-Nickel Steel-Clad Plate, Sheet, and Strip.
A 480/A480M-99b	Specification for General Requirements for Flat- Rolled Stainless and Heat Resisting Steel Plate, Sheet, and Strip.
A666-00	Specification for Annealed or Cold-Worked Austenitic Stainless Steel Sheet, Strip Plate, and Flat Bar.
A693-93 (1999)	Specification for Precipitation-Hardening Stainless and Heat- Resisting Steel Plate, Sheet, and Strip.
Non-Ferrous Sheet and Strip	
B36/B36M-1	Specification for Brass Plate, Sheet, Strip, and Rolled Bar.
B96/B96M-1	Specification for Copper-Silicon Alloy Plate, Sheet, Strip, and Rolled Bar.
B103/B103M-98e2	Specification for Phosphor Bronze Plate, Sheet, Strip, and Rolled Bar.
B121/B121M-01	Specification for Leaded Brass Plate, Sheet, Strip, and Rolled Bar.
B122/B122M-01	Specification for Copper-Nickel-Tin Alloy, Copper Nickel-Zinc Alloy (Nickel-Silver), and Copper-Nickel Plate, Sheet, Strip, and Rolled Bar.
B169M-96	Specification for Aluminum Bronze Sheet, Strip, and Rolled Bar (metric).
B194-01	Specification for Copper-Beryllium Alloy Plate, Sheet, Strip, and Rolled Bar
B422-99	Specification for Copper-Aluminum-Silicon-Cobalt Alloy, Copper-Nickel-Silicon-Magnesium Alloy, Sheet, and Strip.
B465-01e	Specification for Copper-Iron Alloy Plate, Sheet, Strip, and Rolled Bar.
B534-01	Specification for Copper-Cobalt-Beryllium Alloy and Copper-Nickel-Beryllium Alloy Plate, Sheet, and Strip.
B591-98a	Specification for Copper-Zinc-Tin and Copper-Zinc-Iron-Nickel Alloy Plate, Sheet, Strip, and Rolled Bar.
B747-02	Specification for Copper-Zirconium Alloy Sheet and Strip.
B768-99	Specification for Copper-Cobalt-Beryllium Alloy and Copper-Nickel-Beryllium Alloy Sheet and Strip.

Table A3.2 Nominal thickness of sheet and strip

GAUGE	Nominal Thickness			
	Non-Ferrous Material		Ferrous Material	
	Inches	Millimeters	Inches	Millimeters
3	.2294	5.827	.2391	6.073
4	.2043	5.189	.2242	5.695
5	.1819	4.620	.2092	5.314
6	.1620	4.115	.1943	4.935

7	.1443	3.665	.1793	4.554
8	.1285	3.264	.1644	4.176
9	.1019	2.906	.1495	3.797
10	.1019	2.588	.1345	3.416
11	.0907	2.304	.1196	3.030
12	.0808	2.052	.1046	2.657
13	.0720	1.829	.0897	2.278
14	.0641	1.628	.0747	1.897
15	.0571	1.450	.0673	1.709
16	.0508	1.290	.0598	1.519
17	.0453	1.151	.0538	1.367
18	.0403	1.024	.0478	1.214
19	.0359	0.912	.0418	1.062
20	.0320	0.813	.0359	0.912
21	.0285	0.724	.0329	0.836
22	.0253	0.643	.0299	0.759
23	.0226	0.574	.0269	0.683
24	.0201	0.511	.0239	0.607
25	.0179	0.455	.0209	0.531
26	.0159	0.404	.0179	0.455
27	.0142	0.361	.0164	0.417
28	.0126	0.320	.0149	0.378
29	.0113	0.287	.0135	0.343
30	.0100	0.254	.0120	0.305
31	.0089	0.226	.0105	0.267
32	.0080	0.203	.0097	0.246
33	.0071	0.180	.0090	0.229
34	.0063	0.160	.0082	0.208
35	.0056	0.142	.0075	0.191
36	.0050	0.127	.0067	0.170
37	.0045	0.114	.0064	0.162
38	.0040	0.101	.0060	0.152

A3.2 INTERNATIONAL SYSTEM OF UNITS (SI)*

The International System of Units, abbreviated SI (from the French *Le Systeme International d'Unites*), is the modernized version of the metric system established by international agreement. The metric system of measurement was developed during the French Revolution and was first promoted in the United States by Thomas

Jefferson. In 1902, proposed congressional legislation requiring the U.S. Government to use the metric system exclusively was defeated by a single vote.

The SI was established in 1960 by the 11ᵗʰ General Conference on Weights and Measures (CGPM). SI provides a logical and interconnected framework for all measurements in science and industry. The metric system is much simpler to use than the English system because all its units of measurements are divisible by 10.

A3.2.1 SI Base Units

The SI is founded on seven *SI base Units* for seven *base quantities* assumed to be mutually independent, as given in Table A3.3

Table A3.3 SI base units

Base quantity	SI base unit		Definitions
	Name	**Symbol**	
length	meter	m	The meter is the length of the path traveled by light in a vacuum during a time interval of 1/299 792 458 a second.
mass	kilogram	kg	The kilogram is the unit of mass; it is equal to the mass of the international prototype of the kilogram.
time	second	s	The second is the duration of 9 192 631 770 periods of the radiation corresponding to the transition between the two hyperfine levels of the ground state of the cesium 133 atom.
electric current	ampere	A	The ampere is that constant current which, if maintained in two straight parallel conductors of infinite length, of negligible circular cross-section, and placed 1 meter apart in vacuum, would produce between these conductors a force equal to 2×10^{-7} newton per meter of length.
thermodynamic temperature	kelvin	K	The kelvin, the unit of thermodynamic temperature, is the fraction 1/273.16 of the thermodynamic temperature of the triple point of water.
amount of substance	mole	mol	1. The mole is the amount of substance of a system that contains as many elementary entities as there are atoms in 0.012 kilogram of carbon 12; its symbol is "mol."
luminous intensity	candela	cd	2. When the mole is used, the elementary entities must be specified and may be atoms, molecules, ions, electrons, other particles, or specified groups of such particles.
			The candela is the luminous intensity, in a given direction, of a source that emits monochromatic radiation of frequency 540×10^{12} Hertz and that has a radiant intensity in that direction of 1/683 watt per steradian.

A3.2.2 SI Derived Units

Other quantities, called derived quantities, are derived in terms of the seven base quantities via a system of quantity equations. The *SI derived units* for these derived quantities are obtained from these equations and the seven SI base units. Examples of such SI derived units are given in Table A3.4, where it should be noted that the symbol 1 for quantities of dimension 1, such as mass fraction, is generally omitted.

Table A3.4 Examples of SI derived units

Derived quantity	SI derived unit	
	Name	Symbol
area	square meter	m
volume	cubic meter	m^3
speed, velocity	meter per second	m/s
acceleration	meter per second squared	m/s^2
wave number	reciprocal meter	m^{-1}
mass density	kilogram per cubic meter	kg/m^3
specific volume	cubic meter per kilogram	m^3/kg
current density	ampere per square meter	A/m^2
magnetic field strength	ampere per meter	A/m
amount-of-substance concentration	mole per cubic meter	mol/m^3
luminance	candela per square meter	cd/m^2
mass fraction	kilogram per kilogram, which may be represented by the number 1	kg/kg = 1

For ease of understanding and convenience, 22 SI derived units have been given special names and symbols, as shown in Table A3.5.

Table A3.5 SI derived units with special names and symbols

Derived quantity	SI derived unit			
	Name	Symbol	Expression in terms of other SI units	Expression in terms of SI base units
plane angle	radian[a]	rad	–	$m \cdot m^{-1} = 1$[b]
solid angle	steradian[a]	sr[c]	–	$m^2 \cdot m^{-2} = 1$[b]
frequency	hertz	Hz	–	s^{-1}
force	newton	N	–	$m \cdot kg \cdot s^{-2}$
pressure, stress	pascal	Pa	N/m^2	$m^{-1} \cdot kg \cdot s^{-2}$
energy, work, quantity of heat	joule	J	$N \cdot m$	$m^2 \cdot kg \cdot s^{-2}$

power, radiant flux	watt	W	J/s	$m^2 \cdot kg \cdot s^{-3}$
electric charge, quantity of electricity	coulomb	C	–	$s \cdot A$
electric potential difference, electromotive force	volt	V	W/A	$m^2 \cdot kg \cdot s^{-3} \cdot A^{-1}$
capacitance	farad	F	C/V	$m^{-2} \cdot kg^{-1} \cdot s^4 \cdot A^2$
electric resistance	ohm	Ω	V/A	$m^2 \cdot kg \cdot s^{-3} \cdot A^{-2}$
electric conductance	siemens	S	A/V	$m^{-2} \cdot kg^{-1} \cdot s^3 \cdot A^2$
magnetic flux	weber	Wb	V.s	$m^2 \cdot kg \cdot s^{-2} \cdot A^{-1}$
magnetic flux density	tesla	T	Wb/m²	$kg \cdot s^{-2} \cdot A^{-1}$
inductance	henry	H	Wb/A	$m^2 \cdot kg \cdot s^{-2} \cdot A^{-2}$
Celsius temperature	degree Celsius	°C	–	K
luminous flux	lumen	lm	cd·sr[c]	$m^2 \cdot m^{-2} \cdot cd = cd$
illuminance	lux	lx	lm/m²	$m^2 \cdot m^{-4} \cdot cd = m^{-2} \cdot cd$
activity (of a radionuclide)	becquerel	Bq	–	s^{-1}
absorbed dose, specific energy (imparted), kerma	gray	Gy	J/kg	$m^2 \cdot s^{-2}$
dose equivalent[d]	sievert	Sv	J/kg	$m^2 \cdot s^{-2}$
catalytic activity	katal	kat	–	$s^{-1} \cdot mol$

[a] The radian and steradian may be used advantageously in expressions for derived units to distinguish between quantities of a different nature but of the same dimension; some examples are given in Table A3.4.

[b] In practice, the symbols rad and sr are used where appropriate, but the derived unit "1" is generally omitted.

[c] In photometry, the unit name steradian and the unit symbol sr are usually retained in expressions for derived units.

[d] Other quantities expressed in sieverts are ambient dose equivalent, directional dose equivalent, personal dose equivalent, and organ equivalent dose.

Note on degree Celsius. The derived unit in Table 3 with the special name degree Celsius and special symbol °C deserves comment. Because of the way temperature scales used to be defined, it remains common practice to express a thermodynamic temperature, symbol T, in terms of its difference from the reference temperature $T_0 = 273.15$ K, the ice point. This temperature difference is called a Celsius temperature, symbol t, and is defined by the quantity equation

$$t = T - T_0$$

The unit of Celsius temperature is the degree Celsius, symbol °C. The numerical value of a Celsius temperature t expressed in degrees Celsius is given by

$$t = \left(\frac{T}{K} - 273.15 \right) {}^0C$$

It follows from the definition of t that the degree Celsius is equal in magnitude to the kelvin, which in turn implies that the numerical value of a given temperature difference or temperature interval whose value is expressed in the unit degree Celsius (°C) is equal to the numerical value of the same difference or interval when its value is expressed in the unit kelvin (K). Thus, temperature differences or temperature intervals may be expressed in either the degree Celsius or the kelvin using the same numerical value. For example, the Celsius temperature Δt and the thermodynamic temperature ΔT between the melting point of gallium and the triple point of water may be written as

$$\Delta t = 29.7546 \ °C = \Delta T = 29.7546 \ K$$

The special names and symbols of the 22 SI derived units with special names and symbols given in Table A3.5 may themselves be included in the names and symbols of other SI derived units, as shown in Table A3.6.

Table A3.6 Examples of SI derived units whose names and symbols include SI derived units with special names and symbols

Derived quantity	SI derived unit	
	Name	**Symbol**
dynamic viscosity	pascal second	Pa·s
moment of force	newton meter	N·m
surface tension	newton per meter	N/m
angular velocity	radian per second	rad/s
angular acceleration	radian per second squared	rad/s^2
heat flux density, irradiance	watt per square meter	W/m^2
heat capacity, entropy	joule per kelvin	J/K
specific heat capacity, specific entropy	joule per kilogram kelvin	J/(kg·K)
thermal conductivity	watt per meter kelvin	W/(m·K)
energy density	joule per cubic meter	J/m^3
electric field strength	volt per meter	V/m
electric charge density	coulomb per cubic meter	C/m^3
electric flux density	coulomb per square meter	C/m^2
permittivity	farad per meter	F/m
permeability	henry per meter	H/m
molar energy	joule per mole	J/mol
molar entropy, molar heat capacity	joule per mole kelvin	J/(mol·K)
exposure (x and c rays)	coulomb per kilogram	C/kg
absorbed dose rate	gray per second	Gy/s
radiant intensity	watt per steradian	W/sr
Radiance	watt per square meter steradian	W/(m^2·sr)
catalytic (activity) concentration	katal per cubic meter	kat/m^3

A3.2.3 SI Prefixes

The 20 SI prefixes used to form decimal multiples and submultiples of SI units are given in Table A3.7.

It is important to note that the kilogram is the only SI unit with a prefix as part of its name and symbol. Because multiple prefixes may not be used, the prefix names of Table 5 are used for the kilogram with the unit name "gram" and the prefix symbols are used with the unit symbol "g." With this exception, any SI prefix may be used with any SI unit, including the degree Celsius and its symbol °C.

Table A3.7 SI prefixes

Factor	Name	Symbol	Factor	Name	Symbol
10^{24}	yotta	Y	10^{-1}	Deci	d
10^{21}	zetta	Z	10^{-2}	Centi	c
10^{18}	exa	E	10^{-3}	Milli	m
10^{15}	peta	P	10^{-6}	Micro	μ
10^{12}	tera	T	10^{-9}	Nano	n
10^{9}	giga	G	10^{-12}	Pico	p
10^{6}	mega	M	10^{-15}	Femto	f
10^{3}	kilo	k	10^{-18}	Atto	a
10^{2}	hecto	h	10^{-21}	Zepto	z
10^{1}	deka	da	10^{-24}	Yocto	y

A3.2.4 Units outside the SI

Certain units are not part of the International System of Units; that is, they are outside the SI, but are important and widely used. Consistent with the recommendations of the International Committee for Weights and Measures (CIPM, *Comité International des Poids et Mesures*), some of the units in this category that are accepted for use with the SI are given in Table A3.8

Table A3.8 Units outside the SI that are accepted for use with the SI

Name	Symbol	Value in SI units
minute(time)	Min	1m in = 60s
hour	H	1h = 60 min = 3600 s
day	D	1d = 24h = 86 400 s
degree (angle)	°	$1° = (p/180)$ rad
minute (angle)	′	$1 = (1/60)° = (p/10\ 800)$ rad
second (angle)	″	$1 = (1/60) = (p/648\ 000)$ ra d
liter	L	$1\ L = 1\ dm^3 = 10^{-3}\ m^3$
metric ton[a]	T1	$t = 10^3$ kg

[a] In many countries, this unit is called "tonne."

A3.3 METRIC STYLE AND CONVERSIONS

a) Capitals

Units: The names of all units start with a lower case letter except, of course, at the beginning of the sentence. There is one exception: in "degree Celsius" (symbol °C), the unit "degree" is lower case but the modifier "Celsius" is capitalized. Thus, body temperature is written as 37 degrees Celsius.

Symbols: Units' symbols are written in lower case letters except for liter and those units derived from the name of a person (m for meter, but W for watt, Pa for pascal, etc.).

Prefixes: Symbols for prefixes that mean a million or more are capitalized and those less than a million are lower case (M for mega-millions, m for milli-thousandths).

b) Plurals

Units: Names of units are made plural only when the numerical value that precedes them is more than one — for example, 0.25 liter or ¼ liter, but milliliters. Zero degrees Celsius is an exception to this rule.

Symbols: Symbols for units are never pluralized (250 mm = 250 millimeters).

c) Spacing

A space is used between the number and the symbol to which it refers — for example, 7 m, 37.4 kg, 37 °C.

When a metric value is used as a one-thought modifier before a noun, hyphenating the quantity is not necessary. However, if a hyphen is used, write out the name of the metric quality with the numeral and the quantity. For example:

- a 2-liter bottle, NOT a 2-L bottle
- a 100-meter relay, NOT a 100-m relay
- 35-millimeter film, NOT 35-mm film

Spaces are not used between prefixes and unit names, or between symbols and unit symbols. Examples: milligram, mg; kilometer, km.

d) Period

DO NOT use a period with metric unit names and symbols except at the end of a sentence.

e) Decimal Point

The dot or period is used as the decimal point within numbers. In numbers less than one, zero should be written before the decimal point. Examples: 7.038 g; 0.038 g.

f) Conversions

Conversions should follow a rule of reason: do not use more significant digits than are justified by the precision of the original data. For example, 36 inches should be converted to 91 centimeters, not 91.44 centimeters

(36 inches × 2.54 centimeters per inch = 91.44 centimeters), and 40.1 inches converts to 101.9 centimeters, not 101.854. Table A3.9 lists many commonly used conversion factors

Table A3.9 Metric conversion factors (approximate)

Symbol	When You Know Number of	Multiply By	To Find Number of	Symbol
Length				
in	inches	2.54 (exact)	centimeters	cm
ft	feet	0.3	meters	m
yd	yards	0.9	meters	m
mi	miles	1.6	kilometers	km
Area				
in^2	square inches	6.5	square centimeters	cm^2
ft^2	square feet	0.09	square meters	m^2
yd^2	square yards	0.8	square meters	m^2
mi^2	square miles	6.5	square kilometers	km^2
	acres	0.4	hectares	ha
Weight (mass)				
oz	ounces	28	grams	g
lb	pounds	0.45	kilograms	kg
	short tons (2000) pounds)	09	metric tons	t
Volume				
tsp	teaspoon	5	milliliters	mL
Tbsp	tablespoon	15	milliliters	mL
in^3	cubic inches	16	milliliters	mL
fl oz	fluid ounces	30	milliliters	mL
c	cups	0.24	liters	L
pt	pint	0.47	liters	L
qt	quarts	0.95	liters	L
gal	gallons	3.8	liters	L
ft^3	cubic feet	0.03	cubic meters	m^3
yd^3	cubic yards	0.76	cubic meters	m^3
Pressure				
inHg	inches of mercury	3.4	kilopascals	kPa
psi	pounds per square inch	6.9	kilopascals	kPa
Temperature (exact)				
^0F	degrees Fahrenheit	$\frac{5}{9}\left(^0F - 32\right)$	degrees Celsius	^0C

*Information by courtesy of National Institute of Standards and Technology (NIST)

Appendix
Four

Technical Specification of the
Helical and Belleville Springs

A spring is defined as an elastic body, whose function is to distort when loaded and to recover its original shape when load is removed. The various important applications of springs in die design are as follows:

- To apply forces as in strippers and ejectors
- To control motion by maintaining contact between two elements, as in elastic stop pins
- To cushion absorb or control energy as in guide post systems or elastic guide rail systems

Although there are many types of the springs, two types of springs are most often used in die design and manufacturing: helical and Belleville springs.

A4.1 HELICAL SPRINGS

Helical springs are made up of a wire coiled in the form of a helix; they are intended primarily for compression or tension loads. The cross-section of the wire from which the spring is made is circular, square, or rectangular. As the load increases, the number of inactive coils also increases due to the seating of the end coils; the amount of increase varies from 0.5 to 1 turn. At the usual die working loads, the end connections for helical springs form a 3/4 coil on each end of spring. So, the total number of turns is $i = n + 1.5$.

Table A4.1 provides technical specifications for cold-formed helical springs of the circular cross-section wire that is most often used as a die part.

Table A4.1 Technical specification for helical springs*

d (mm)	D (mm)	F_{max} (N)	$i = 6.5$		$i = 8.5$		$i = 10.5$		$i = 13.5$		$i = 17.5$	
			L (mm)	f_{max} (mm)	L (mm)	f_{max} (mm)	L (mm)	f_{max} (mm)	L (mm)	f_{max} (mm)	L (mm)	f_{max} (mm)
0.5	3.5	15	6.4	3.1	8.3	4.0	10.9	5.6	14.3	7.5	18.1	9.3
	4.0	13	7.6	4.3	10.3	6.0	13.0	7.7	17.1	10.3	22.5	13.7
	4.5	11	8.7	5.4	11.9	7.6	15.0	9.7	19.8	13.0	26.2	17.4
	5.5	9.0	11.9	8.6	16.5	12.2	21.0	15.6	27.6	20.8	36.6	27.8
0.63	4.2	24	7.5	3.4	10.1	4.7	12.7	6.1	16.8	8.1	21.7	10.7
	4.7	22	8.4	4.3	11.4	6.0	14.3	7.7	18.8	10.3	24.7	13.7
	5.7	17	10.5	6.4	14.3	8.9	18.0	11.4	23.7	15.2	31.3	20.3
	6.7	14	13.3	9.2	18.3	12.9	23.3	16.6	30.5	22.0	40.2	29.2
0.8	5.1	41	9.0	3.8	12.1	5.3	15.3	6.9	20.4	9.2	26.2	12.2
	6.1	33	10.9	5.3	14.8	8.0	18.7	10.3	24.5	13.7	30.4	18.4
	7.1	28	13.5	8.0	18.4	11.6	25.7	17.3	30.6	19.6	40.4	26.4
	8.1	24	15.7	10.7	21.8	15.0	27.8	19.4	36.6	25.8	48.4	34.5
1.0	6.5	62	11.4	4.9	15.4	6.9	19.4	8.9	25.3	11.8	33.3	15.8
	7.5	53	13.5	7.0	18.3	9.8	23.1	12.6	30.3	16.8	40.4	22.4
	8.5	46	15.8	9.3	21.5	13.0	27.2	16.7	35.8	22.4	47.1	29.6
	10.5	36	21.1	14.6	28.5	20.4	36.9	26.4	48.5	35.0	64.5	47.0
1.25	8.2	91	14.3	6.1	19.2	8.6	24.3	11.1	31.6	14.7	41.5	19.6
	9.2	80	16.2	8.0	22.0	11.3	27.7	14.5	36.2	19.3	47.6	25.7
	11.2	64	20.7	12.5	28.2	17.7	35.8	22.6	48.9	32.0	61.9	40.0
	13.2	53	26.2	18.0	35.9	25.3	45.6	32.4	59.9	43.0	79.4	57.5
1.6	9.9	152	16.8	6.4	22.6	9.0	28.3	11.5	37.0	15.4	48.4	20.4
	11.9	123	20.2	9.8	27.2	13.7	34.5	17.7	45.2	23.6	59.4	31.4
	13.9	103	24.3	14.0	33.0	19.4	41.5	25.0	55.0	34.0	27.5	44.5
	15.9	88	29.1	18.7	39.8	26.2	50.6	33.8	66.5	45.0	88.0	60.0
2.0	12.7	218	21.0	8.0	28.2	11.2	35.4	14.4	46.3	19.3	60.6	25.6
	14.7	184	24.2	11.2	32.7	15.7	41.5	20.5	54.0	27.0	71.0	36.0
	16.7	159	28.3	15.3	38.4	21.4	48.5	27.5	63.7	36.7	84.0	49.0
	20.7	125	37.3	24.3	51.0	34.0	64.8	43.8	85.4	58.4	113.0	77.0

2.5	15.7	320	25.4	9.1	34.0	12.7	42.6	16.3	55.8	21.8	73.0	29.0
	17.7	280	28.1	11.8	38.0	16.6	47.6	21.3	62.4	28.4	82.0	38.0
	21.7	220	35.8	19.5	48.6	27.3	61.3	35.0	80.8	46.8	106.0	62.0
	25.7	180	43.9	27.6	60.0	38.7	76.0	49.6	100.5	66.5	132.5	88.5
3.2	19.1	530	30.5	9.7	40.8	13.6	51.5	17.5	66.5	23.2	87.0	31.0
	23.1	420	35.9	15.1	48.5	21.2	61.0	27.2	79.4	36.4	104.5	48.5
	27.1	350	43.0	22.2	58.4	31.0	74.0	39.8	96.0	53.0	127.0	71.0
	32.1	290	52.6	31.8	71.8	44.5	91.4	57.4	119.0	76.5	158.0	102.0
4.0	25.0	740	39.0	12.9	52.0	18.0	65.0	23.2	85.0	31.0	111.0	41.3
	29.0	520	44.5	18.6	60.2	26.2	75.5	33.5	99.0	44.7	130.0	59.5
	34.0	510	52.0	25.9	70.3	36.3	88.5	46.6	116.0	62.2	153.0	83.3
	41.0	420	66.0	39.8	89.8	55.6	113.5	71.5	149.0	95.0	197.0	127.0
5.0	31.0	1080	47.0	14.5	63.5	20.4	98.0	26.2	103.0	34.8	135.0	46.0
	36.0	910	53.5	21.0	72.5	29.4	112.5	37.8	118.5	50.5	155.0	67.0
	43.0	740	64.0	31.2	87.0	43.8	134.0	56.4	143.0	75.0	188.0	100.0
	51.0	610	78.0	45.6	107.0	64.0	166.0	82.0	177.0	109.0	234.0	146.0
6.3	39.1	1650	57.5	17.6	78.7	24.7	123.0	31.8	127.0	42.2	167.0	56.2
	46.1	1360	67.0	26.0	90.5	36.4	139.0	46.6	148.0	62.5	194.0	83.0
	54.1	1130	79.0	37.8	107.0	53.0	164.0	68.8	176.0	90.5	232.0	121.0
	64.1	940	96.5	55.4	132.0	77.8	202.0	100.0	218.0	133.0	288.0	171.0
8.0	49.5	2580	74.0	21.8	98.5	30.4	150.0	39.2	160.0	52.0	210.0	69.5
	75.5	2160	82.5	30.6	111.0	43.0	166.5	55.0	182.0	73.5	238.0	98.0
	67.5	1790	96.5	44.4	130.0	62.0	194.0	80.0	214.0	106.0	282.0	142.0
	80.5	1470	117.0	65.5	160.0	92.0	238.0	118.0	265.0	157.0	350.0	210.0
10.0	62.0	3560	90.0	24.8	120.	34.8	164.0	44.8	119.0	59.0	254.0	79.0
	72.0	2960	99.0	34.0	133.0	47.8	182.5	61.5	217.0	82.0	284.0	109.0
	85.0	2450	114.5	49.5	154.5	69.5	203.5	89.0	254.0	119.0	333.0	158.0
	107.0	2000	139.0	74.0	189.0	104.0	154.0	133.0	312.0	177.0	412.0	237.0
11.0	70.0	4000	98.6	26.6	131.0	37.0	176.0	48.0	213.0	64.0	277.0	85.0
	80.0	3420	106.0	37.0	145.5	51.5	201.0	66.5	238.0	89.0	310.0	118.0
	90.0	3000	120.5	48.5	162.0	68.0	219.0	87.5	265.0	116.0	347.0	155.0
	110.0	2400	148.0	76.0	201.0	107.0	276.0	138.0	332.0	183.0	436.0	244.0
12.5	75.0	5150	107.0	24.8	142.0	34.8	195.0	44.5	229.0	59.5	298.0	79.0
	90.0	4150	120.0	38.2	160.0	53.5	219.0	69.0	261.0	92.0	342.0	123.0
	100.0	3680	130.5	48.5	175.0	68.0	252.0	87.0	285.0	116.0	375.0	155.0
	125.0	2860	162.0	80.0	219.0	112.0	294.0	144.0	361.0	192.0	475.0	256.0

*Source: Dr. Binko Musafija, Obrada Metala Plasticnom Deformcijom.

A4.2 BELLEVILLE SPRINGS

Belleville springs, disc springs, Belleville washers, and conical compression washers are all names for the same type of spring. They have a frustum conical shape, which gives the washer a spring characteristic. Originally developed in the mid-19th century by Julian Belleville, a Belleville spring experiences deflection and stress when a load is applied in the axial direction only. It has a very nonlinear relation between the load applied and the axial deflection. The stress distribution is non-uniform for this spring. The axial force is applied at the edge of the inner diameter, inducing stress at the inner surface and at the outer surface, which depends on geometric parameters. The deflections and the stresses induced at the inner surface and at the outer surface depend on the ratios of height to thickness (h/T) and the ratio of the outer diameter to the inner diameter (D/d).

Belleville springs can be used in four stacking positions:

1. Single, one spring
2. Parallel, all springs stacked the same way
3. Series, all springs stacked opposite each other
4. Parallel-series, a combination of #2 and #3

Belleville springs in stacked arrangements provide increased load and/or deflection.

Single stacking. A single Belleville spring has a specific load and deflection.

Parallel stacking. Stacking two springs in parallel doubles the load of the single spring with no increase in deflection.

Series stacking. Stacking two springs in series doubles the deflection of a single spring with no increase in load.

Parallel-series stacking. A parallel-series combination results in the load of two springs and the deflection of two springs.

The advantages of disc springs compared to other types of springs include the following:

- A wide range of load/deflection characteristics
- High spring loads with small deflections
- Efficient use of space
- Flexibility in stack arrangement to meet application requirements
- Keep bolted joints tight
- Long service life
- Simple adjustment of the load and deflection of a spring stack by adding or removing individual springs
- Self-damping.

The spring geometry consists of four parameters which is shown in Fig. A4.1.

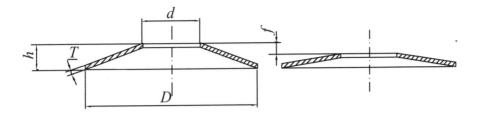

Fig. A4.1 Belleville spring.

Table A4.2 provides technical specifications for Belleville springs that are most often used as a die parts.

Table A4.2 Technical specification for Belleville springs*

DIMENSIONS						DIMENSIONS					
D (mm)	d (mm)	T (mm)	h (mm)	F_{max} (N)	f_{max} (mm)	D (mm)	d (mm)	T (mm)	h (mm)	F_{max} (N)	f_{max} (mm)
8	3.2	0.3	0.55	126	0.25	40	14.3	1.25	2.65	1984	1.40
8	4.2	0.2	0.45	42	0.25	40	14.3	1.5	2.80	3184	1.30
8	4.2	0.4	0.60	269	0.20	40	14.3	2.0	3.05	6096	1.05
10	3.2	0.3	0.65	108	0.35	40	16.3	1.5	2.80	3281	1.30
10	4.2	0.5	0.75	377	0.25	40	16.3	2.0	3.10	6580	1.10
10	5.2	0.4	0.7	257	0.30	40	18.3	2.0	3.15	7171	1.15
10	5.2	0.5	0.75	418	0.25	40	20.4	1.0	2.30	1072	1.30
12	4.2	0.4	0.80	208	0.40	40	20.4	1.5	2.65	3201	1.15
12	5.2	0.5	0.90	424	0.40	40	20.4	2.0	3.10	7258	1.10
12	6.2	0.5	0.85	404	0.35	40	20.4	2.5	3.45	12243	0.95
12	6.2	0.6	0.95	699	0.35	45	22.4	1.25	2.85	2007	1.60
12.5	6.2	0.35	0.80	160	0.45	45	22.4	1.75	3.05	4475	1.30
12.5	6.2	0.5	0.85	363	035	45	22.4	2.5	3.50	10037	1.0
12.5	6.2	0.7	1.00	855	0.30	50	18.4	1.5	3.15	2600	1.65
14	7.2	0.35	0.80	131	0.45	50	18.4	2.0	3.65	6163	1.65
14	7.2	0.5	0.90	338	0.40	50	20.4	2.0	3.50	5745	1.50
14	7.2	0.8	1.10	1040	0.30	50	20.4	2.5	3.85	10098	1.35
15	5.2	0.4	0.95	181	0.55	50	22.4	2.0	3.60	6329	1.60
15	6.2	0.6	1.05	733	0.40	50	22.4	2.5	3.90	10817	1.40
15	8.2	0.7	1.10	844	0.40	50	25.4	1.5	3.10	2844	1.60
15	8.2	0.8	1.20	1261	040	50	25.4	2.0	3.40	5898	1.40
16	8.2	0.4	0.90	165	0.50	50	25.4	2.5	3.90	11519	1.40
16	8.2	0.6	1.05	503	0.45	50	25.4	3.0	4.10	15640	1.10

16	8.2	0.9	1.25	1319	0.35	56	28.5	1.5	3.45	2766	1.95
18	6.2	0.4	1.00	137	0.6	56	28.5	2.0	3.60	5379	1.60
18	6.2	0.5	1.10	267	0.6	56	28.5	3.0	4.30	14752	1.30
18	6.2	0.6	1.20	462	0.6	60	20.5	2.0	4.20	5636	2.20
18	6.2	0.7	1.40	855	0.7	60	25.5	3.0	4.65	15002	1.65
18	6.2	0.8	1.5	1277	0.7	60	30.5	2.5	4.50	11433	2.0
18	8.2	0.7	1.25	725	0.55	60	30.5	3.0	4.70	16792	1.70
18	8.2	0.8	1.30	984	0.50	60	30.5	3.5	5.0	23528	1.50
18	8.2	1.0	1.50	1921	050	63	31	1.8	4.15	4463	2.35
18	9.2	0.7	1.20	223	0.60	63	31	2.5	4.25	8904	1.75
18	9.2	1.0	1.40	699	0.50	63	31	3.0	4.70	14946	1.70
20	8.2	0.6	1.30	453	0.7	63	31	3.5	4.90	19545	1.40
20	8.2	0.7	1.35	668	0.65	70	30.5	2.5	4.90	9360	2.40
20	8.2	0.8	1.40	921	0.60	70	30.5	3.0	5.10	14152	2.10
20	8.2	0.9	1.50	1311	0.60	70	30.5	3.0	5.10	15218	2.10
20	10.2	0.5	1.15	268	0.65	70	35.5	4.0	5.80	30990	1.80
20	10.2	0.8	1.35	929	0.55	71	36	2.0	4.60	5426	2.60
20	10.2	0.9	1.45	1323	0.55	71	36	2.5	4.50	8152	2.00
20	10.2	1.0	1.55	1815	0.55	71	36	4.0	5.60	26712	1.60
23	8.2	0.7	1.50	602	0.8	80	41	2.25	5.20	6950	2.95
23	8.2	0.8	1.55	842	0.75	80	41	3.0	5.30	12844	2.30
23	8.2	0.9	1.70	1279	0.80	80	41	4.0	6.20	29122	2.20
23	10.2	1.0	1.70	1629	0.70	80	41	5.0	6.70	43952	1.70
23	12.2	1.5	2.10	3000	0.60	90	46	2.5	5.70	8157	3.20
25	12.2	0.7	1.60	1050	0.70	90	46	3.5	6.00	17487	2.50
25	12.2	1.5	2.05	3821	0.55	90	46	5.0	7.00	40786	2.00
28	10.2	0.8	1.75	723	0.95	100	41	4.0	7.20	24547	3.20
28	10.2	1.0	2.00	1486	0.95	100	41	5.0	7.75	41201	2.75
28	10.2	1.5	2.20	3511	0.70	100	51	2.7	6.20	9091	3.50
28	12.2	1.0	1.95	1482	0.95	100	51	3.5	6.30	15843	2.80
28	12.2	1.5	2.25	3949	0.75	100	51	4.0	7.00	25338	3.00
28	14.2	0.8	1.80	859	1.0	100	51	5.0	7.80	46189	2.80
28	14.2	1.25	1.10	1342	0.80	100	51	6.0	8.20	62711	2.20
28	14.2	1.5	2.40	3680	0.65	112	57	3.0	6.90	11064	3.90
34	12.3	1.0	2.20	1208	1.20	112	57	4.0	7.20	21518	3.20
34	12.3	1.25	2.45	2359	1.20	112	57	6.0	8.50	56737	2.50
34	12.3	1.5	2.70	4078	1.20	125	64	3.5	8.00	16335	4.50

34	14.3	1.25	2.40	2347	1.15	125	64	5.0	8.50	37041	3.50
34	14.3	1.5	2.55	3704	1.05	140	72	3.8	8.70	18199	4.90
34	16.3	1.5	2.55	3908	1.05	140	72	5.0	9.00	33848	4.00
34	16.3	2.0	2.85	7498	0.85	160	82	6.0	10.50	50260	4.50
35.5	18.3	0.9	2.05	884	1.15	180	92	6.0	11.1	44930	5.10
35.5	18.3	2.0	2.80	6747	0.80	200	102	5.5	12.50	38423	7.00

* Sorce: DIN 2023

GLOSSARY

ADAPTOR. A block used to mount a die to a press slide.

ADVANCE. The amount that a strip or part moves forward in a progressive stamping operation. Also called the "jump."

AGING. A process of change in the mechanical properties of metals and alloys. Aging takes place over a period of time at room temperature, but it occurs more quickly at higher temperatures.

AIR BENDING. A forming operation in which the metal part is formed without the punch striking the bottom of the die. Contact between work material and the tool is made at only three points in the cross section.

AIR BENDING DIE. A die in which the lank is bent without sticking to the bottom of the die. A blank's contact with the die is made at only three lines—the nose line of the punch and two edges of the die opening.

AIR CUSHION. An air cushion actuated by a large pressurized cylinder, located beneath the bed of a press, used to apply upward pressure to the lower die.

AIR HARDENING. The process by which an alloy steel is heated to the proper hardening temperature and then allowed to cool in air.

AIRCRAFT QUALITY. "Aircraft quality" metals and alloys are tested during manufacturing processes and approved as having a quality appropriate for use in aircraft parts.

AISI. American Iron and Steel Institute.

ALCLAD. A very thin layer of highly pure aluminum is used to coat sheets or plates of stronger aluminum alloy.

ALLOTROPISM. Reversible phenomenon of metals existing in more than one crystal structure.

ALLOY STEEL. Steel that is alloyed with significant quantities of chemical elements such as nickel, chromium, molybdenum, or vanadium.

ALLOYING ELEMENT. A chemical element added to a pure metal during melting processes to impart certain physical and mechanical properties.

ALLOYS. Pure metal that has been melted together with other chemical elements into a new metal structure having specific physical and mechanical properties.

AMORPHOUS. Without definite form; non-crystalline; having no crystal structure.

ANISOTROPY. The property of having different qualities when a material is tested in various directions.

ANNEALING. Heating and slowly cooling a metal to remove stresses; this treatment can make the metal softer, refine its structure, change its ductility, or change its strength, toughness, or other qualities.

ASTM. American Society for Testing and Materials.

AUSTENITE STAINLESS STEEL. Non-magnetic stainless steel that cannot be hardened through heat treatment.

AUTOMATIC PRESS STOP. A machine-generated signal for stopping the acting of a press by automatically disengaging the clutch mechanism and engaging the brake mechanism.

BEADING. A sheet-metal forming operation in which a narrow ridge is formed in a workpiece to strengthen it.

BED. Bottom stationary structural member of a press, used as a holder for a die.

BED CUSHION. Commonly required for draw tooling, a system that applies resistance when pushed upon. This resistance can be dynamic or statically controlled throughout the stroke. Bed cushions have a "pusher pin plate" that is located just beneath the bed bolster. The bed bolster is provided with multiple through holes where "pusher pins" are inserted. These pins are used by the tooling to generate resistive force as the hydraulic press ram pushes down.

BED HEIGHT. The distance from the bottom of the hydraulic press structure to the working height or top of the bed bolster.

BEND ANGLE. The angle through which a part is bent.

BEND RADIUS. The inside radius of a bend in a formed section.

BEND TEST. A test to determine the bendability of a given metal by bending a specimen between two fixed points.

BIAXIAL STRESS. The stress experienced by metals in which applied stresses are limited to two principal axes.

BLANK. An unformed piece of sheet metal before the beginning of forming operations; a workpiece resulting from a blanking operation.

BLANK DEVELOPMENT. Determining the best shape and size to make a blank for a specific part.

BLANKHOLDER. The part that holds the sheet metal in place while forming operations such as deep drawing are being performed.

BLANKING. Die cutting of the outside shape of a part.

BOTTOMING. Sheet-metal forming operation in which the punch and the die make complete contact with the workpiece by the use of increased bending force.

BRITTLE. An adjective describing a quality of material in which breakage takes place easily, with little or no bending or other plastic deformation.

BULGING. Expanding the diameter of a tube or other deep-drawn part by pressure from inside.

BURR. A thin, rigid, sharp edge left on sheet metal blanks by cutting operations.

BURNISH. Shiny, smooth area above the breakout on a sheared edge. Also a shiny, smooth surface produced by application of pressure and movement.

BURRING. Operation in which the rough-cut edges of metal are deburred.

CAD. Computer-Aided Design.

CAM. A device used to control the motion of slides of the die components during the press stroke.

CAM DRIVER. A component of the die with an angular surface; transfers vertical motion of the press slide to the mating angular surface on the cam slide.

CAM SLIDE. A device for performing work at an angle to the press stroke.

CARBON STEEL. A steel with up to 1.7% carbon and without any substantial amount of other alloying elements. Also termed plain carbon steel.

CARBURIZING. Adding carbon to the surface of steel by heating it in contact with one of a number of carbon-rich media.

CAST IRON. Iron that contains between 1.7 and 4 percent of carbon.

CLEARANCE. The free space between two mating parts. In closed contours, clearance is measured on one side.

CLEAVAGE. Breakage or planned fracture made by splitting a material to follow a crystallographic plane.

COIL. A length of steel that has been rolled.

COINING. A compressive sheet metal forming operation in which all surfaces of the workpiece are put in contact with the technological surface of the punch and the die by the use of increased forming force.

COLD ROLLING. Passing unheated metal through rollers to reduce ts thickness. Creates a smooth surface with slightly increased skin hardness.

COLD WORKING. Any operation of plastic deformation that is performed at room temperature.

COMPOUND DIE. A die that can perform more than one operation with one press-stroke.

COMPRESSIVE STRENGTH. The ability of a material to withstand compressive loads without being crushed when the material is in compression.

CRANK PRESS. A press in which the ram is powered by a crankshaft.

CREEP. The phenomenon of a material's elongation under tension, over a certain period of time, usually at higher-than-room temperatures.

CRIMPING. Securing a seam or folding part of a sheet over another part by a forming operation.

CROSS-ROLLING. Rolling a sheet at an angle of 90-degrees to the direction that it was previously rolled.

CROWN or CAMBER. The condition when a sheet or roll of metal is contoured so that the thickness or diameter is greater at the center than at the edges.

CRYSTAL. In solid materials, the repetitive structure in which atoms are arranged.

CRYSTAL LATTICE. The pattern of arrangement of atoms in a crystalline structure.

CUPPING. Producing a cup-shaped part from a sheet-metal blank by means of a forming operation.

DAYLIGHT. The maximum distance between the bed and the upper, movable, pressing surfaces of a press.

DEBURRING. Removing burrs from metal by various procedures.

DENDRITE. A branched or leaf-like crystalline structure, usually occurring during the solidification of metals.

DIE. a) In a general sense, an entire press tool with all components taken together; b) In a more limited sense, a component that is machined to receive the blank, as differentiated from the "punch," which is its opposite member.

DIE CUSHION. Pressurized cylinder which is used to apply upward pressure to the ejector of a part or stripper.

DIFFUSION. The movement of atoms or molecules across mating metal surfaces or within the material itself.

DILATATION. A change in volume or dimension.

DOUBLE-ACTION PRESS. A press that performs two parallel movements independently; the machine may have an interior slide to form the part and an exterior slide for the blank holder.

DOWEL. A rigid pin, case hardened, that fits into a corresponding hole to align two die parts.

DRAW RADIUS. The radius at the edge of a die or punch over which sheet metal is drawn.

DRAW RING. A circular-shaped part of the die that is used in a deep drawing operation to control material flow and wrinkling.

DRAWABILITY. The ability of a sheet metal to be formed, or drawn, into a cupped or cavity shape without cracking or otherwise failing.

DRAWING QUALITY STEEL. Flat-rolled steel sheet that can undergo forming (especially deep drawing) without defects.

DRIVER.	A component of the die assembly with one or more angular surfaces that apply force by the vertical movement of the press to mating angular surfaces on a cam slide.
DUCTILITY.	The ability of a material to be stretched under the application of tensile load and still retain the new shape when the load is removed.
EARING.	The formation of wavy edge projections around the top edge of a deep-drawn part.
ECCENTRIC PRESS.	A machine that exerts working pressure, using an eccentric shaft.
EJECTOR.	A mechanism operated by mechanical, hydraulic, or pneumatic power mechanism for removing a workpiece from a die.
ELASTICITY.	The property of a material that allows it to deform under load and immediately return to its original size and shape after the load is removed.
ELASTOMER.	A rubber-like synthetic polymer such as silicone rubber.
ELONGATION.	The amount of permanent extension of the material before it fractures.
ENGINEERING STRAIN.	*See Strain.*
ENGINEERING STRESS.	*See Stress.*
FEED.	The precise linear travel of the stock strip at each press stroke, equal to the interstation distance. Also called pitch.
FIXTURE.	A device to locate and hold a workpiece or components in position.
FLANGE.	The formed rim of a part, generally designed for stiffening and fastening.
FLOATING DIE.	A die that is mounted so that it can tolerate lateral or vertical movement during the forming process.
FOIL.	Extremely thin sheet or strip of metal.
FORMABILITY.	The capacity of a material to undergo plastic deformation without fracture.
FORMING LIMIT DIAGRAM (FLD).	In order to assess the formability of sheet metals while forming a workpiece, circle grind analyses is used to construct a forming limit diagram of the material to be used.
FRACTURE.	The surface appearance of a freshly broken material.
FRACTURE STRESS.	Nominal stress at fracture.
GRAIN.	A single crystalline structure within a microstructure. Examples are polycrystalline metals or alloys.
GRAIN BOUNDARY.	The interface or meeting plane at the boundaries of grains.
GRAIN DIRECTION.	Crystalline orientation of material in the direction of mill-rolling.
GRAIN GROWTH.	An increase in the size of the grain, usually occurring when a metal is heated.
GRAPHITE.	Carbon; the hexagonal crystalline structure is layered.

HALF SHEARING. Partial penetration by piercing, making a locating button with a height no more than half the material thickness.

HARDENABILITY. The ability of steel that determines the ease of transformation of austenite to another structure when cooled from hardening temperature.

HARDENING. The process of heating and cooling or quenching a metal to increase its hardness.

HARDNESS. The ability of a metal to resist indentation, scratching, and surface abrasion by another hard object.

HEMMING (Flattening). The process or bending or folding sheet metal over itself to produce a smooth border.

HEAT-RESISTANT STEEL. Steels that are made to be used at high temperatures; they keep most of their strength and are able to resist rust or oxidation under high-temperature conditions.

HEAT TREATMENT. Process of heating and cooling of a metal (or alloy) to obtain certain chosen qualities.

HOT ROLLED STEEL. Steel that has been roller-formed into a sheet or another shape from a hot plastic state.

HOT WORKING. The process of forming metals at temperatures higher than their recrystallization point.

HYDRAULIC PRESS. Machine that exerts working pressure by hydraulic means.

INCLUSIONS. A solid or gaseous foreign substance encased in liquid metal during solidification.

INSERT. A part of a die or mold made to be removable.

IRONING. A deep-drawing operation accomplished by reducing the wall thickness and outside diameter of a drawn cup.

ISO. International Standards Organization.

ISOTROPIC. Having physical and mechanical properties of material that are the same regardless of the direction of measurement.

KILLED STEEL. Deoxidized steel.

KNOCKOUT. A mechanism for freeing a workpiece from a die.

LATTICE. The arrangement of atoms in a given crystalline structure

LEVELING. The process of flattening a sheet or strip that has been rolled.

LIMIT DRAW RATIO (LDR). The greatest ratio of a blank diameter to a punch diameter that can be cup-drawn without cracking or producing other defects in the workpiece.

LOCATING PIN. Also called a pilot pin.

LUBRICANT. A substance used to reduce friction.

MACROSTRUCTURE. Crystal structure that can be viewed when magnified from 1 to 10X.

MALLEABILITY. The capability of material to be deformed permanently into various shapes, under the application of a compressive load, without breaking.

MASS PRODUCTION. The production of parts in large quantities, usually over 100,000 per year.

METAL THINNING. The normal reduction in thickness that takes place during metal-forming operations.

MICROSTRUCTURE. Crystalline structure that can be seen at magnifications greater than 10X.

MILD STEEL. Carbon steel that has no more than 0.25 percent carbon.

MODULUS OF ELASTICITY. The ratio of stress to strain in the elastic domain in tension or compression.

MODULUS OF RESILIENCE. Area under the elastic portion of a stress-strain curve, representing the energy that can be absorbed without permanent deformation.

MODULUS OF RIGIDITY. The ratio of shear stress to shear strain in the elastic range.

MONOCRYSTAL. Single crystal.

MULTIPLE DIE. A die for the production of two or more identical parts at one press stroke.

NATURAL STRAIN. True strain.

NECKING. A reduction in the cross-section of a localized area of a part that has been subjected to tensile stress during the process of plastic deformation.

NOMINAL STRAIN. Engineering strain.

NOMINAL STRESS. Engineering stress.

NORMAL ANISOTROPY. Anisotropy normal to the plane of a sheet or plate.

NORMALIZING. The process of heating a metal to about 50 degrees centigrade above the critical temperature and cooling again, in still air, to room temperature.

NOTCHING. Punching operation in which the punch removes material from the edge of a strip or blank.

OIL QUENCH. Operation in which the material is quenched from the hardening temperature in oil as the cooling medium.

ORANGE PEEL. Surface condition appearance on metals; bears a resemblance to orange skin texture due to the coarse grain size after forming.

PARTING LINE. The plane of two mating surfaces of a die or mold.

PATTERN DIRECTION. Orientation of features of surface patterns on sheet and coils.

PENETRATION. Depth of a cutting operation before breakout occurs.

PICKLING. The process of removing surface oxides by exposure to chemical or electrochemical reaction.

PICKUP. The inadvertent transfer of metal from a workpiece to the surface of the die.

PIERCING. Cutting or punching of openings such as holes and slots in material.

PILOT. A pin for locating a workpiece in a die from a previously punched hole.

PITCH NOTCH. A notch usually cut on one side of strip in a die to control stock width and progression of the strip. Also called a French notch.

PLANAR ANISOTROPY. Anisotropy in the plane of a sheet or plate.

PLASTIC DEFORMATION. Permanent change in the shape of the work material.

PLASTICITY. The property of a material that permits it to undergo a permanent change in shape without cracking.

PRESSURE PAD. A lower die component that is used to hold work material while the part is being formed.

PRESSWORKING. General sheet-forming operations performed in a press.

PROGRESSIVE DIE. A die that can perform two or more operations sequentially with one stroke of the press.

PUCKERING. Formation of small folds or wrinkles in a cup in a deep-drawing operation.

PUNCH. The die component attached to the die set that forms or cuts the workpiece.

PUNCH FACE. Striking end of punch.

PUNCH HOLDER. A die component used to mount the punch usually on the upper die shoe.

QUENCH CRACKING. Appearance of cracks in a metal piece during the process of quenching from a high temperature.

QUENCHING. The quick cooling of a heated workpiece in water or oil.

QUECK-CHANGE INSERT. Tool sections or die components that be changed without removing the entire die from the press.

RAM. A moving member of a press to which a die or punch is attached.

RD. Rolling direction.

RECOVERY. Removal or reduction of the effects of cold working without motion of grain boundaries.

RECRYSTALLIZATION. Recrystallization occurs when new equiaxial strain-free crystal grains form after a metal has been heated to or above its recrystallization temperature.

REDRAWING. The second and successive deep drawing operation in which a workpiece becomes deeper and a reduced in its cross-sectional dimensions.

REDUCTION.	Change in a material's thickness or cross-section in the process of plastic deformation.
RESIDUAL STRESS.	Any stress remaining in a body after external forces have been removed.
RESIN.	Natural or synthetic polymer that has been neither filled nor reinforced.
RUBBER FORMING.	A sheet metal forming operation in which a rubber cushion block is used as a functional male or female die part.
SECONDARY OPERATIONS.	Treatments performed after the initial metal stamping process. Example: cleaning, heat treating , or deburring.
SEGMENT DIE.	A die made of the parts that can be separated for ready removal from the workpiece. Also known as a split die.
SHAVING.	Removal of a thin layer of material with a sharp edge of die or punch.
SHEAR MODULUS.	Modulus of rigidity.
SHEAR STRENGTH.	The ability of a material to withstand offset or transverse loads without rupture occurring.
SINGLE ACTION.	A press utilizing only one moving element.
SLAG.	The scrap that results from a punching operation.
SLIP.	Plastic deformation by shear along a crystallographic plane.
SPRINGBACK.	Elastic rebounding of formed material, as in bending.
STOCK GUIDE.	A device used to direct a strip or sheet material through the die.
STOP BLOCK.	A die component used to act as a reference point for the height of the tool set up.
STOP PIN.	A die component used to direct a strip or sheet material through the die.
STRAIN (Engineering).	The ratio of the change in dimension from the original dimension as a result of the application of stresses in working.
STRAIN (True).	The natural logarithm of the ratio of the final to the original dimensions.
STRAIN ENERGY.	The amount of energy used in the process of plastic deformation.
STRESS (Engineering).	The ratio of the load to the original cross-sectional area.
STRESS (True).	The ratio of load to the instantaneous cross-sectional area.
STRINGER.	Elongated inclusions or impurities, such as oxides and sulfides, which appear in worked metals. Generally, these impurities occur in the same direction as that of the plastic deformation.
STRIPPER.	A die component designed to surround a punch that strips the scrap from the punch.
STRIPS.	Sheet metal sheared into long, narrow pieces.
STROKE.	Ram travel from top dead center to bottom dead center.
STRUCTURAL DAMAGE.	Damage or defects that occur in a material as a result of plastic deformation.

SUBGRAIN. A part of a grain that is not lined up in the same direction as the neighboring grains.

SUPERPLASTICITY. Unusual capability of a material to withstand large uniform strains before it necks and fails.

TEMPERING. Reheating to a temperature below the critical range after a steel has been quenched; tempering is done to relieve quenching stresses or to develop desired strength characteristics.

TENSILE STRENGTH. The greatest longitudinal stress that a material can sustain without breaking.

TOOL STEEL. Any high carbon or alloy steel capable of being suitable tempered for use in tool-and-die manufacturing.

TOOLING HOLE. Holes provided in a workpiece for locating purposes during production.

TOUGHNESS. The ability of a material to withstand shatter. A material that shatters easily is brittle.

TRANSFER DIE. Progressive die in which the workpiece is transferred from station to station by a mechanical system.

TRIMMING. Adjusting the size of a workpiece by removing small amounts of excess metal.

TRIPLE ACTION PRESS. A press having three moving slides—two slides moving in the same direction and a third lower slide moving upward through the fixed press bed.

ULTIMATE TENSILE STRENGTH. The greatest value of engineering stress determined by a tension test.

UNIFORM STRAIN. Strain experienced by a tensile test bar before necking.

UNIT CELL. The smallest unit in a repetitive crystalline structure.

UTS. Ultimate tensile strength.

VACANCY. A blank space that should be occupied by an atom in a crystalline structure.

VENT. An opening, usually quite small, to allow the escape of gases from a mold.

WARM WORKING. Working a metal at a temperature between the ambient temperature and the recrystallization temperature; neither hot nor cold working.

WOOD'S METAL. A specific alloy made of lead, tin, bismuth, and cadmium.

WORK HARDENING. Increase in tensile strength of metal resulting from cold work.

YIELD STRESSES. Stress at which the material yields and begins to deform plastically.

YOUNG'S MODULUS. Modulus of elasticity.

BIBLIOGRAPHY

Alting, L. *Manufacturing Engineering Processes*, 2nd ed. New York: Marcel Dekker, Inc. 1994.

Arnold, J. *Die MakersHandbook*. New York: Industrial Press, Inc. 2000.

Boljanovic,V. *Some Results and Prospectives toward Future Research on Bending Process in Elastic Dies.* Belgrade: Fifth Yugoslav Aerocosmonautic Congress, vol. 1, 981.

Boljanovic,V. *Influence of Bending Angles on the Value of Springback.* Mostar: Sixteenth Conference of Yugoslav Manufacturing Engineering, 1982.

Boljanovic,V. *Research of Influential Factors and Their Correlative Links in Aluminum Alloys by Using Methods Depending on Means of Work Applied.* Mostar: Dzemal Bijedic University, 1982.

Boljanovic, V. *Influence of Additional Reduction of Sheet Metal Thickness on the Shape Steadiness of an Element Formed by Bending.* Ph.D. dissertation. Mostar: Dzemal Bijedic University, 1989.

Boljanovic, V. *Die Design Fundamentals,* 3rd ed. New York: Industrial Press, 2005.

Boljanovic, V. *Metal Shaping Process.* New York: Industrial Press, 2009.

Boljanovic, V. *Sheet Metal Stamping Dies — Die Design and Die Making Practice.* New York: Industrial Press, 2012.

Courtney, T.H. *Mechanical Behavior of Materials.* New York: McGraw-Hall, 1990.

Groover, M.P.F. *Fundamentals of Modern Manufacturing.* New Jersey: Prentice Hall, 1996.

Kalpakjian, S., and S.R. Schmid. *Manufacturing Engineering and Technology*, 5th ed. New Jersey: Pearson Education, Inc., 2006.

Kalpakjian, S. *Manufacturing Processes for Engineering Materials.* Reading, Massachusetts: Addison-Wesley, 1985.

Kuhtarov, V. I. *Stampi Dlja Holodnoj Listovoj Stampovki.* Moscow: Masgiz, 1960.

Lisov, M. N. *Teorija i Rascet Processov Izgatovlenija Detalej Metodom Gibki.* Moscow: Masinostroenie, 1967.

Malov, E.N.*Technologia Holodnoi Stampovki.* Moskow: Oborongiz, 1963.

Makelt, H. *Die Mechanishen Pressen.* Munich: Carl Hanser Verlag, 1961.

Mosnin, E.N. *Gibka i Pravka na Rotacionnih Masinah.* Moscow: Masinostroenie, 1967.

Musafija, B. *Obrada Metala Plasticnom Deformacijom.* Sarajevo: Svjetlost, 1988.

Neferov, A.P. *Konstruirovanie i Izgotovlenie Stampov.* Moskow: Masinostroenie, 1973.

Oberg E, F.D. Jones, H. L. Horbot., H.H Ryffel. *Machinery's Handbook* 29th ed. New York: Industrial Press, 2012.

Oehler G. *Schnitt, Stanz und Zienhwerkzeuge.* Berlin: Springer Verlag, 1966.

Pollack, W.H. *Tool Design.* Reston: Reston Publishing Company, Inc., 1976.

Popov, E.A. *Osnovi Teorii Stampovki.* Moscow: Masinostroenie, 1968.

Romanowski, W. P. *Handbuchder Stanzereitechnik,* Berlin: VEB Verlag Technik, 1959.

Smith, D., *Die Design Handbook,* 3rd ed. Dearborn: Society of Manufacturing Engineers, 1990.

Yankee, H.W. *Manufacturing Processes.* Englewood Cliffs: Prentice Hall, Inc., 1979.

Zubcov, M.E. *Listovaja Stampovka.* Leningrad: Masinostroenie, 1961.

INDEX